Lecture Notes in Earth Sciences   58

W0043960

Springer-Verlag Berlin Heidelberg GmbH

Golam Sarwar
Gerald M. Friedman

Post-Devonian Sediment Cover over New York State

Evidence from Fluid-Inclusion, Organic Maturation, Clay Diagenesis and Stable Isotope Studies

 Springer

Authors

Dr. Golam Sarwar
Northeastern Science Foundation affiliated with
Brooklyn College, City University of New York
P.O. Box 746, 15 Third Street, Troy, NY 12181, USA

Prof. Dr. Gerald M. Friedman
Department of Geology, Brooklyn College
and Graduate Center, City University of New York
and Northeastern Science Foundation affiliated with Brooklyn College
Rensselaer Center of Applied Biology
P.O. Box 746, 15 Third Street, Troy, NY 12181, USA

Library of Congress Cataloging-in-Publication Data

Sarwar, Golam.
    Thick post-Devonian sediment cover over New York State : evidence
from fluid-inclusion, organic maturation, clay diagenesis, and
stable isotope studies / Golam Sarwar, Gerald M. Friedman.
      p.   cm. -- (Lecture notes in earth sciences ; 58)
    Includes bibliographical references.
    ISBN 978-3-540-59458-1   ISBN 978-3-540-49271-9 (eBook)
    DOI 10.1007/978-3-540-49271-9
    1. Paleothermometry--New York (State) 2. Sedimentation and
deposition--New York (State) 3. Geology, Stratigraphic.
4. Geology--New York (State)   I. Friedman, Gerald M. II. Title.
III. Series.
QE721.2.P3S27 1995
551.7'09747--dc20   95-19708
                                                               CIP

"For all Lecture Notes in Earth Sciences published till now please see final pages of the book"

ISBN 978-3-540-59458-1

© Springer-Verlag Berlin Heidelberg 1995
Originally published by Springer-Verlag Berlin Heidelberg New York in 1995

Typesetting: Camera ready by authors
SPIN: 10503636   32/3142-543210 - Printed on acid-free paper

PREFACE

In the geologic record, vertical crustal uplift has often resulted in erosional removal of huge thicknesses of sedimentary strata. If the uplift is of a broad regional nature or the uplifted strata remain relatively undeformed and sediments deposited after the uplift are not preserved, the magnitude of uplift and subsequent erosion may be difficult to quantify. This may lead to misinterpretation or omission of chapters of geologic history of a region. Fortunately, a number of indirect methods can be used to infer the thicknesses of missing strata and reconstruct the geologic history.

Our book titled "Thick Post-Devonian Sediment Cover Over New York State: Evidence from Fluid-Inclusion, Organic Maturation, Clay Diagenesis and Stable Isotope Studies" uses four techniques of paleotemperature measurements in sedimentary rocks in order to determine burial depths of the existing Paleozoic strata in New York State. Since every technique has its own analytical and interpretative uncertainties, the use of four techniques allowed us to place a better constraint on our results. We show how regionally extensive paleotemperature data can be used to estimate the thicknesses of strata lost from an uplifted sedimentary basin. We also provide a tentative tectonic-, paleogeographic- and depositional history of New York State after the Devonian when the missing strata were deposited.

The work described in this book was a part of doctoral research of Golam Sarwar under Gerald M. Friedman at the Department of Earth and Environmental Sciences, the City University of New York. The laboratory work was carried out at Brooklyn College, New York, and the Northeastern Science Foundation, Troy, affiliated with Brooklyn College. Financial support for Sarwar was received from a PSC-CUNY grant (#669198) held by Friedman, a Student Honorarium Award from the New York State Geological Survey, and a stable-isotope analysis grant from Geochem Laboratories, Cambridge, Massachusetts. We are grateful to the New York State Geologic Survey for not only providing financial support, but also for supplying core samples, well logs, and other pertinent information. Many people assisted us through the development of this work. They include Dr. Robert Fakundiny, Prof. Somdev Bhattacharji, Prof. Nicholas Coch, Dr. D. W. Fisher, Dr. William Kelley, Dr. Hiroshi Homa, Dr. Ali Kaya, and Dr. Jon Bass. The final manuscript improved considerably owing to critical review by Dr. Jon Bass. To all of them we give our grateful thanks.

CONTENTS

THICK POST-DEVONIAN SEDIMENT COVER OVER NEW YORK STATE: EVIDENCE FROM FLUID-INCLUSION, ORGANIC MATURATION, CLAY DIAGENESIS AND STABLE ISOTOPE STUDIES

Golam Sarwar[1] and Gerald M. Friedman[2]

[1] Northeastern Science Foundation affiliated with Brooklyn College of the City University of New York, P.O. Box 746, 15 Third street, Troy, New York, 12181
[2] Department of Geology, Brooklyn College and Graduate Center of the City University of New York, Brooklyn, NY 11210 and the Northeastern Science Foundation affiliated with Brooklyn College, P.O. Box 746, 15 Third Street, Troy, NY 12181-0764

ABSTRACT

Paleothermic study of surface and shallow subsurface samples from sixteen Paleozoic rock units of New York State using the methods of fluid inclusions, organic maturation, clay diagenesis, and stable isotopes indicate paleotemperatures often exceeding 200° C. The high paleotemperatures are inferred to have been attained from normal geothermal heat during former deep burial. Calculated burial depths indicate that substantial thicknesses of post-Devonian strata were once present across the state reaching more than 6km in southeastern and west-central New York. From various geologic considerations it appears that sedimentation in New York State continued to the end of the Permian. It is hypothesized that the primary sources of post-Devonian sediments were the Late Pennsylvanian - Early Permian Alleghanian thrust sheets rapidly emplaced in the vicinity of southeastern New York and the moderate uplift along the Findlay-Algonquin Arch northwest of New York flexurally uplifted in response to the Alleghanian thrust load. The Alleghanian movement, unlike in much of southern and central Appalachian basin where it led to deformation and uplift, marked an interval of sedimentation in New York and probably the adjoining areas. Sedimentation took place in an inland basin in the form of alluvial-fan, braid-plain, flood-plain, lacustrine, and small deltaic deposits. The subsequent uplift and erosion of the area was probably initiated by the pre-rifting doming of eastern North America as it migrated over hot-spots or mantle plumes during the Triassic period.

1 INTRODUCTION

The post-Devonian geologic history of the northern Appalachian basin has been a subject of controversy since it was suggested in Epstein et al.'s (1977) and Harris' (1979) works that the surficial rocks in the northern Appalachian basin, including those in New York State, have been exhumed from great depths after removal of strata, much of which might have been post-Devonian in age.

Except for the Cenozoic cover of Long Island and parts of Staten Island, and the Triassic rocks of the Newark Basin in the southeastern corner of the state, the rocks exposed in New York range from Precambrian to Upper Devonian in age. Mainly on the basis of the absence of post-Devonian strata in New York and thinning of Carboniferous strata from Pennsylvania toward New York it has been traditionally believed that since the end-Devonian little sediment had accumulated in New York (see Meckel, 1970, for example) and the region has largely witnessed a continuous cycle of erosion.

However, many recent studies, using techniques as diverse as fluid inclusions, stable isotopes, clay diagenesis, organic maturation, fission tracks, and K-Ar dating have found that the rocks currently exposed at the surface or at shallow depths in New York State and the vicinity had experienced high paleotemperatures (Crough, 1981; Friedman and Sanders, 1982, 1983; Lakatos and Miller, 1983; Johnsson, 1986; Friedman, 1987a, 1987b; Jackson et al. 1988). According to most of these studies, the high paleotemperatures measured in these rocks were attained during deep burial, and to account for the burial it became necessary to infer that much of the eroded strata was post-Devonian in age.

A great many implications arise from the results of these studies. If sedimentation indeed continued long after the Devonian in New York, as many of these studies suggest, this will require revision of much of the present concepts of the region's post-Devonian geologic history, and questions like how long after the Devonian did sedimentation continue, what were the depositional environments, what was the paleogeography like, where did the additional sediments come from, what happened to these sediments, etc. have to be addressed. Because of the limited scope of most of the previous studies, these questions remain largely unanswered. In the present broad-based study, samples from as many as sixteen rock units of New York State have been analyzed for paleotemperature signatures by using four principal techniques, namely, fluid inclusions, organic maturation, clay diagenesis, and stable isotopes. Through analytical and interpretive methods described below, the spatial variation in the amount of missing strata in New York has been estimated. An attempt has been made to reconstruct the possible depositional setting and paleogeography of post-Devonian New York.

2 GEOLOGICAL SETTING

Much of New York State lies within the northern part of the Appalachian basin of eastern North America (fig. 1). The Appalachian basin is a multi-stage foreland basin located between the Appalachian orogen to the east and the stable interior provinces to the west. Its eastern and southeastern boundaries are marked by the crystalline core of the Appalachian orogen (the Piedmont Province), and western and northwestern margins by the Cincinnati Arch and the Findlay-Algonquin Arch, respectively. These arches, which are basement-involved structural highs, separate the Appalachian basin from the intra-cratonic Illinois and Michigan basins. North of New York State, the basin becomes greatly constricted, and in northwest Newfoundland it tapers off between the Laurentian Shield and the so called Dunnage Terrane (fig. 2).

The distribution of bedrocks in New York is shown in the simplified geologic map of figure 3. The oldest rocks in New York are the Precambrian metamorphic rocks of the roughly circular Adirondack Highlands to the north and the narrow ridge of the Hudson Highlands in the southeast. These rocks were originally deposited as sedimentary and volcanic rocks in shallow seas about 1.3 billion years ago and were later buried to depths of as much as 30 km, intensely deformed, intruded by magma, and metamorphosed during the Grenville orogeny which peaked between 1.1 and 1.05 billion years ago (Whitney, 1991). These Grenvillian basement rocks underlie all of New York and much of eastern North America, and the Adirondacks and Hudson Highlands are among several scattered uplifts in which these rocks are now exposed.

A long cycle of erosion, lasting several hundred million years, followed the Grenville orogeny when as much as 25 km of rock was removed from New York (Whitney, 1991). By the late Precambrian time the Grenvillian basement began rifting which finally led to the creation of an Atlantic- type passive margin along the entire length of eastern North America (Price and Hatcher, 1983; Tankard, 1986). On this late Precambrian passive margin, syn-rift clastic sediments were deposited and are probably represented in the southern and central Appalachians by the coarse clastics of the Ocoee Group ("Great Smoky Supergroup" in Price and Hatcher, 1983) in the Great Smoky Mountains of Tennessee and North Carolina and the Glenarm Group of Maryland and Pennsylvania (Rodgers, 1967). In New York area, Precambrian clastics are absent, suggesting that the rifting and opening of the Proto-Atlantic or Iapetus Ocean took place later in the northern Appalachian region.

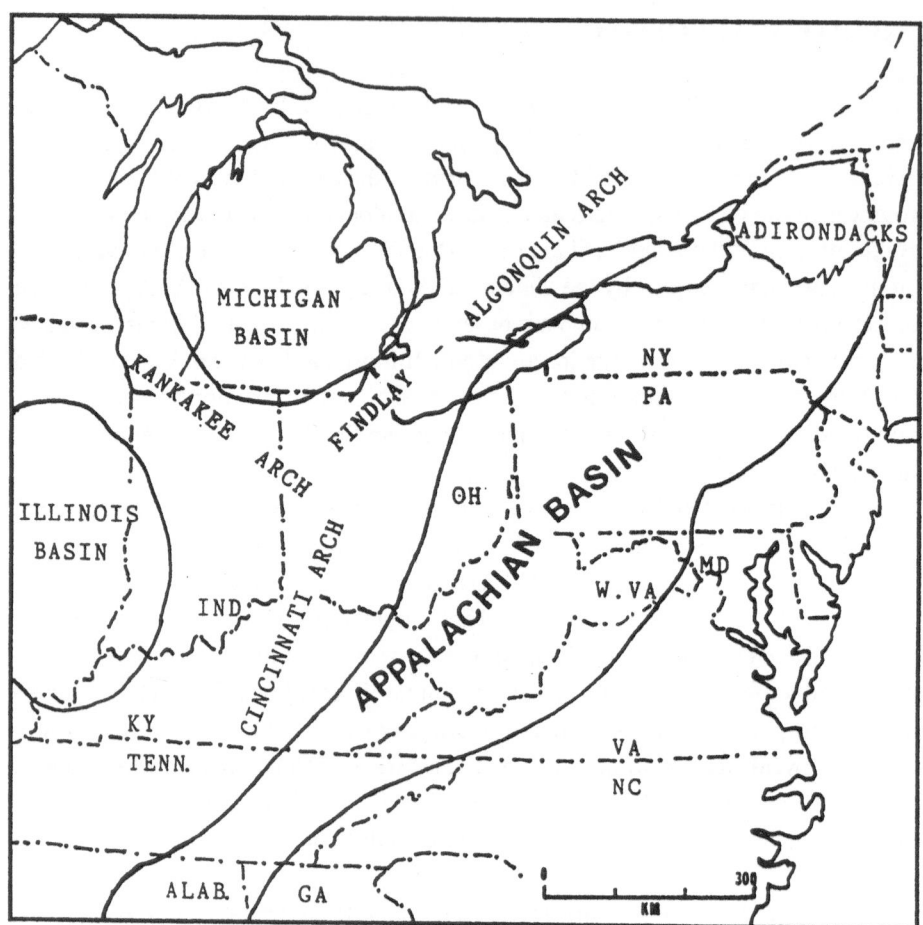

Figure 1: Map showing the location of the Appalachian foreland basin relative to the interior basins and arches of eastern North America. Abbreviations: NY = New York, PA = Pennsylvania, W. VA = West Virginia, MD = Maryland, OH = Ohio, VA = Virginia, NC = North Carolina, GA = Georgia, KY = Kentucky, Tenn. = Tennessee, Ind. = Indiana, Alab. = Alabama. (Modified from Tankard, 1986)

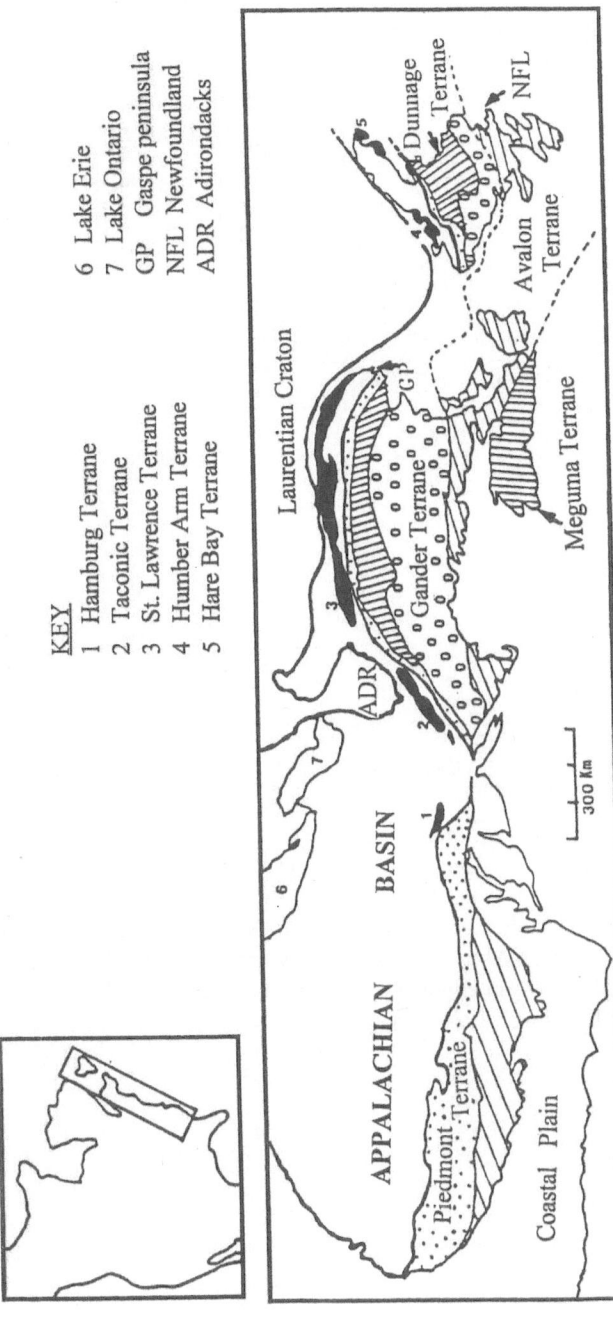

Figure 2: Map showing the position of the Appalachian foreland basin with respect to the suspect terranes of the Appalachian orogen. (Modified from Williams and Hatcher, 1982, and Faill, 1985)

KEY
1 Hamburg Terrane
2 Taconic Terrane
3 St. Lawrence Terrane
4 Humber Arm Terrane
5 Hare Bay Terrane

6 Lake Erie
7 Lake Ontario
GP Gaspe peninsula
NFL Newfoundland
ADR Adirondacks

Laurentian Craton

Dunnage Terrane

NFL

Avalon Terrane

Meguma Terrane

GP

Gander Terrane

ADR

APPALACHIAN BASIN

Piedmont Terrane

Coastal Plain

300 Km

6

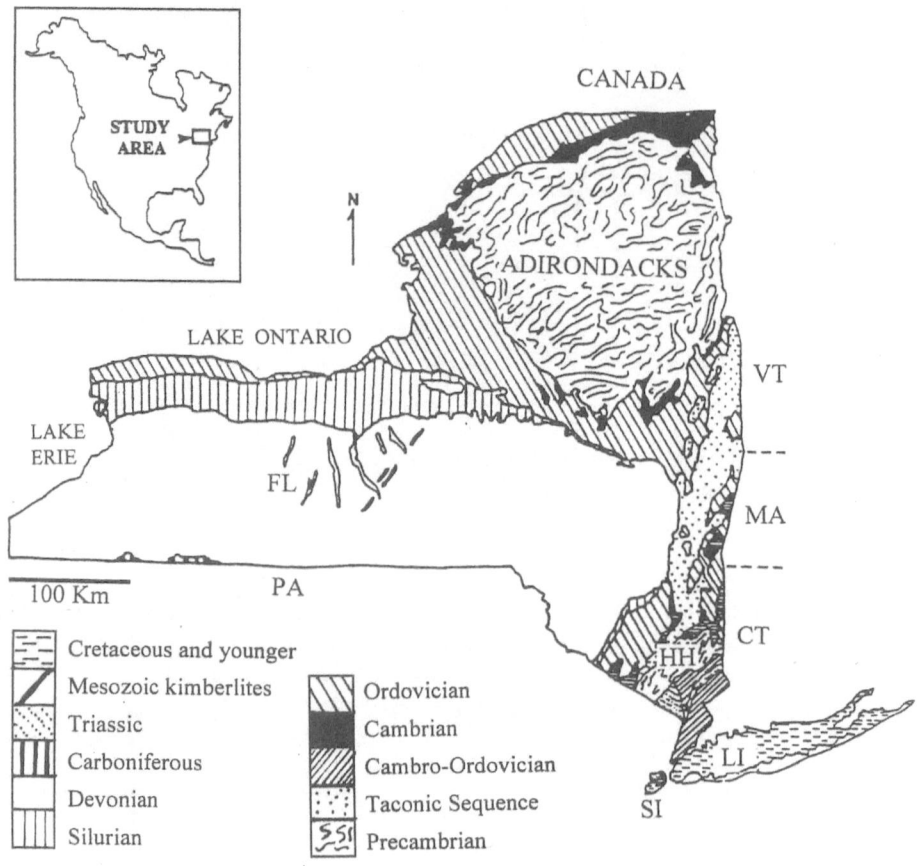

Figure 3: Geologic map of New York. (Modified from New York State Geological Survey map in Geogram, 1990 series). Abbreviations: VT = Vermont, MA = Massachusetts, CT = Connecticut, HH = Hudson Highlands, LI = Long Island, SI = Staten Island, PA = Pennsylvania, FL = Finger Lakes.

The rocks of early Cambrian to early Ordovician age in New York were deposited in a shelf environment in the west and a slope and rise environment in the east. A shallow shelf existed over much of New York, as far east as the northern flanks of the Hudson Highlands (Broughton et al. 1966; Landing, 1991). The shelf sequence begins with transgressive deposits consisting of clean quartzite or sandstones - the Early Cambrian Poughquag Formation (of the Wappinger Group) in southeastern New York and basal Potsdam Formation of the Beekmantown Group elsewhere in New York (table 1). Deposition of these basal clastics, which had their source in the cratonic lowlands to the west, were followed by extensive carbonate sedimentation. In southeastern New York, the carbonates of the Wappinger Group (Cambro-Ordovician) were deposited while in central and western New York the carbonates of the Beekmantown Group were deposited.

While deposition of carbonates was going on throughout New York State, farther east deposition of both coarse and fine grained detrital sediments was taking place in a slope-rise environment (Keith and Friedman, 1977; Friedman et al. 1982). These sediments, spanning the early Cambrian to early Middle Ordovician interval, are represented by muddy sandstones (Rensselaer, Bomosen) and quartzite (Mud Pond, Diamond Rock) of the Taconic allochthonous sequence (fig. 3, table 1)) east of the Hudson River. These older rocks of the slope-rise sequence become progressively thicker and coarser grained to the west indicating a western provenance, probably from the Laurentian lowlands (Fakundiny et al. 1989).

2.1 Taconic orogeny and following events

Passive margin deposition was terminated by the Taconic tectonic event which began in the early Middle Ordovician and lasted through the early Silurian. The Taconic event involved collision of the North American continental margin with a belt of oceanic island arcs above an east-southeast dipping subduction zone (Williams, 1979; Rowley and Kidd, 1981; Hiscott et al. 1986).

As the continental margin approached the subduction zone, the seaward part of the carbonate shelf floundered, probably due to "normal faulting caused by plate flexure with downbending" (Hiscott et al. 1986). As convergence continued, stacked thrust-sheets presumably overrode the outer parts of the continental terrace and resulted in rapid flexural downwarping of the ancestral continental slope and rise (Price and Hatcher, 1983; Quinlan and Beaumont, 1984; Tankard, 1986). The downwarped continental margin bounded on the oceanic side by the rising Taconic orogenic belt created the elongate Appalachian foreland basin between Newfoundland, Canada in the north and Alabama in the south.

Table 1: Generalized stratigraphic column for different areas of New York State (adapted from Rickard, 1975 and Fisher, 1977).

PERIOD	WESTERN NEW YORK	CENTRAL NEW YORK	EASTERN NEW YORK	
LATE CRETACEOUS			Raritan Formation	
EARLY JURASSIC - LATE TRIASSIC			Newark Group	
EARLY PENNSYLVANIAN	Olean Formation			
EARLY MISSISSIPIAN	Knapp Formation			
UPPER DEVONIAN	Connewango Group Conneaut Group Canadaway Group West Falls Group Sonyea Group Genesee Group	West Falls Group Sonyea Group Genesee Group	Slide Mountain Fm. Walton Formation Delware River Fm. Oneonta Formation	
MIDDLE DEVONIAN	Tully Formation Hamilton Group Onondaga Fm.	Tully Formation Hamilton Group Onondaga Fm.	Gilboa Formation Hamilton Group Onondaga Fm.	
LOWER DEVONIAN	Tristates Group	Tristates Group Helderberg Group	Tristates Group Helderberg Group	
UPPER SILURIAN	Rondout Formation Salina Group Lockport Group	Rondout Formation Salina Group Lockport Group	Rondout Formation Salina Group Bloomsburg Fm.	
LOWER SILURIAN	Clinton Group Medina Group	Clinton Group Medina Group	Shawangunk Fm.	
UPPER ORDOVICIAN	Queenston Fm. Lorraine Group Utica Formation Trenton Group Black River Group	Queenston Fm. Lorraine Group Utica Formation Trenton Group Black River Group	Frankfurt/Quassaic Fm. Utica Formation/ Snake Hill Formation	
LOWER ORDOVICIAN			Beekmantown Group/ Wappinger Group	
UPPER CAMBRIAN	Beekmantown Group	Beekmantown Group	Wappinger / Stockbridge Groups	TACONIC SEQUENCE
MIDDLE - LOWER CAMBRIAN			Wappinger / Stockbridge Groups	
PRECAMBRIAN	Grenville Basement	Grenville Basement	Grenville Basement	

In the deepest part of this young foreland basin, close to the orogenic belt, argillaceous limestones with slide sheets were deposited first, followed by dark graptolitic shale and flysch derived from the orogenic belt (Hiscott et al. 1984). So, for the first time the source of clastic sediments had shifted from west to east. In New York, east of the Hudson River, and in western New England these deep-water flysch sediments are represented by the Middle Ordovician Normanskill Group of the Taconic allochthon. The rocks of the Normanskill Group coarsen and thicken to the east and also contain volcanic ash, metamorphic rock fragments and "exotic sand-sized grains of chromite" indicating that the sediments were derived from volcanoes and upthrusted oceanic sediments of the accretionary prism to the east (Stevens, 1970; Fakundiny et al. 1989).

While the eastern part of the foreland basin, proximal to the orogenic belt, deepened under thrust-loads, the western, landward part probably became the locus of a shifting "peripheral bulge" (Quinlan and Beaumont, 1984; Tankard, 1986). Peripheral bulging or upwarping is believed to result from visco-elastic adjustment of the lithosphere to marginal thrust-loading. If the load remains in place for a long period of time(during a period of tectonic quiescence, for example), the proximal part of the basin subsides dragging the forebulge in toward the thrust-load (fig. 4).

Peripheral upwarping due to Taconic thrust-loading is believed to have caused widespread erosion of the late Cambrian and early Ordovician carbonates of the ancestral passive margin and is represented by the Knox-Beekmantown unconformity in New York and elsewhere in the Appalachian basin (Quinlan and Beaumont, 1984; Tankard, 1986). Jacobi (1981) has also hypothesized the origin of this unconformity with the aid of a "forebulge", but the proposed mechanism of erosion differs in detail from that of Quinlan and Beaumont (1984).

The Knox-Beekmantown unconformity also coincides with creation of the sub-Tippecanoe unconformity, one of the six major unconfomities recognized by Sloss (1963, 1972) on the North American craton as well as in Russia and Europe, attributed to a global, eustatic sea level drop. Therefore, basin margin thrust-loading and "peripheral upwarping" (Quinlan and Beaumont, 1984) may not have been the only, or the primary, cause of erosion that resulted in the Knox-Beekmantown unconformity in the Appalachian basin. Perhaps thrust-load induced upwarping and eustatic sea-level drop occurred simultaneously in response to major plate readjustments and this enhanced the rate of erosion.

The stratigraphic position and tectonic relationships within the Taconic allochthon (fig. 3, table 1) has long been a source of controversy which is still not fully resolved (Friedman et al. 1982, p. 35 - 46). The allochthon consists of Lower Cambrian to pre-Upper Ordovician clastic sequences that contain the slope-rise sediments of the pre-Taconic passive margin as well as the deep-water flysch

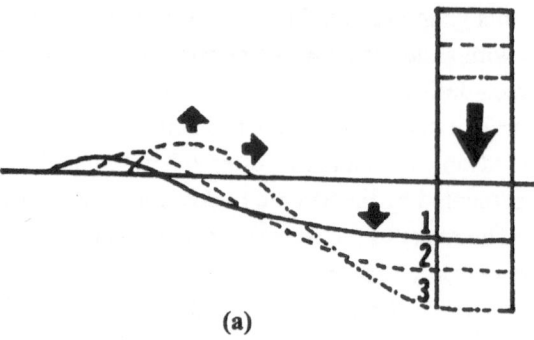

(a)

1. Overthrust loading - flexural deformation

2. Relaxation phase - viscoelastic response

3. Renewed overthrust loading - flexural deformation

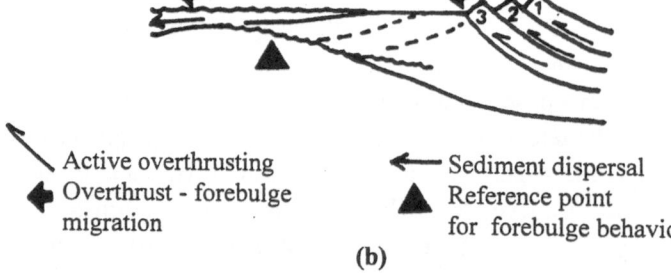

Active overthrusting   Sediment dispersal
Overthrust - forebulge   ▲ Reference point
migration   for forebulge behavior

(b)

Figure 4: Foreland deformation model applied by various authors to the Appalachian basin. (a) Stages (1, 2, 3) of deformation of a visco-elastic plate under an applied load. As the load remains stationary for a long period of time, visco-elastic relaxation causes uplift of the forebulge and its migration toward the load. (B) Stages (1, 2, 3) of deformation in a foreland basin in response to thrust-loading at the basin margin. Thrust-loading causes uplift of the forebulge on the platform side of the basin and its migration toward the thrust-margin may cause widespread erosion (stage - 2) in the basin. (After Quinlan and Beaumont, 1984 and Tankard, 1986).

deposits (Normanskill Group) of the foreland basin. The allochthon is believed to have been transported 150 to 200km from the east and thrust upon the largely coeval shelf carbonates and clastics to the west (Friedman et al. 1982; Stanley and Ratcliffe, 1985). There is considerable debate concerning the nature of the lower contact of the allochthon, which is rarely exposed. Some workers propose that the Taconic allochthon was thrust upon the synorogenic flysch sediments rather than on shelf carbonates (Zen, 1967, 1972; Bosworth et al. 1988).

The former extent of the Taconic allochthon to the west of its present outcrops is not known. There are some indications that it was upthrusted on the eastern edge of the Adirondacks, and might have extended an undetermined distance to the west (Whitney and Davin, 1987).

In southeastern New York, the Taconic allochthon is overlain by a late Middle Ordovician (Mohawkian) "autochthonous" terrigenous sequence (Snake Hill shale and Quassaic sandstones and conglomerate). These rocks record the progressive basin filling and westward progradation of the terrigenous detritus through time. Elsewhere in New York, the shallow-marine carbonates of the Black River Group (table 1) were being deposited when terrigenous sedimentation began in southeastern New York. The Taconic-derived sediments (the Utica Shale and sandstones and shales of the Lorraine Group) steadily encroached upon the shelf carbonates (Black River and overlying Trenton Group) and finally buried them as far west as the present shores of Lake Ontario (Fisher, 1977).

The end of the Ordovician in the northern and central Appalachian basin was marked by deltaic sedimentation. An enormous delta, known as the Queenston Delta, prograded westward west to the midcontinent (Broughton et al. 1966; Rickard, 1991a). In western New York the Lorraine Group rocks are conformably overlain by the non-marine deltaic strata of the Queenston Formation. In the eastern part of New York, however, subsequent erosion of coeval deposits is represented by a prominent angular unconformity between tilted Middle Ordovician Snake Hill shales and flat-lying, Silurian Shawangunk Conglomerate. About 800m of the Upper Ordovician section might have been lost from eastern New York (Fakundiny et al. 1989). The erosion and deformation of Ordovician rocks in the eastern part of New York suggests, that, toward the end of the Ordovician, Taconic deformation progressed from the east into New York State. The deformation was accompanied by block faulting in eastern and southeastern Adirondacks as well as the lower Hudson Valley, and ultrabasic intrusions (Cortlandt Mafic Complex) and metamorphism (Bedford Gneiss) in the Manhattan Prong. The extensive Taconic deformations and magmatic events of New England and eastern Canadian seaboard have been summarized by Rodgers (1967, 1971). In Central and Southern Appalachian, Taconic deformation was relatively mild.

The late Ordovician disturbance in New York was probably related to the arrival of the "Gander microcontinent" (fig. 2) that was accreted east of an earlier "Dunnage microcontinent", along the northeastern margin of the orogen (Williams and Hatcher, 1982). The collision with the Gander microcontinent probably caused a major uplift of the Taconics and caused shedding of enormous detritus from which the Queenston delta formed.

The Silurian, although the shortest of the Paleozoic periods, was marked by the most diverse depositional environments in New York region. When the Silurian period began, probably the entire state of New York was dry land with much of the state covered by red mud-flats of the Queenston delta (Broughton et al. 1966; Rickard, 1991a). A brief incursion of the sea took place in western New York in the early Silurian, but was soon halted by renewed tectonic activity in the Taconics which initiated another, smaller pulse of clastics to spread across the state and is now preserved in the deltaic sediments of the Medina Group.

Following the deposition and reworking of the Medina strata, the Silurian shoreline gradually advanced eastward due to a rise in relative sea level. The Silurian sea was relatively shallow, and fluctuated from well-oxygenated to anoxic conditions, probably in response to frequent eustatic changes or epeirogenic earth movement. In this sea the faunally and lithologically diverse sediments (sandstones; black, green and red shales; limestones; conglomerates; ironstones) of the Clinton Group were deposited (Gillette, 1947; Rickard, 1975; Rickard, 1991a).

During the deposition of the Clinton Group sediments in central and western New York, more than 300m of quartz sand and pebbles of the Shawangunk Formation were deposited in southeastern New York and Pennsylvania in alluvial fans and braided streams. The fluvial sediments of the Shawangunk Formation were the products of the rapidly eroding Taconic Mountains to the east where tectonism had ceased in the early Silurian (Epstein and Lyttle, 1987).

The Taconic Mountains served as the major sediment source until the end of the Lower Silurian when the Rochester Shale was deposited. The Rochester Shale represents the final stages of the erosional decay of the orogen. This faunally diverse marine shale was deposited over much of New York and Ontario, Canada. In the west and northwest, the crest of the Algonquin Arch served as a high-energy shoreline along which a carbonate platform, dominated by crinoidal shoals, was maintained while deposition of mud (Rochester Shale) took place farther east (Tankard, 1986).

Following Clinton, the carbonates of the Lockport Group were deposited in the western half of New York and in the Ontario Peninsula of Canada in a shallow, clear and warm sea. The earliest coral reefs of New York are found in the Lockport dolostones. In east-central New York the Lockport carbonates grade into shale, and in southeastern New York at this time the fluvial sandstones and conglomerates of the upper Shawangunk Formation were deposited.

In late Silurian, the climate in the New York region turned arid and the sea withdrew to the west of New York leaving behind hypersaline embayments and lagoons on a vast sabkha-type mud flat (Treesh and Friedman, 1974). In this depositional setting, gypsiferous shales and thick salt beds of the Salina Group were deposited in central and western New York. In the southeast, the Bloomsburg and Poxono Island shales were deposited at this time, but no evaporites are known (Rickard, 1975). The lithologic change from conglomerate and sandstones of the Shawangunk Formation to shales of the Bloomsburg and Poxono Island formations probably reflects progressively deeper erosion into older argillaceous source rocks in the Taconics.

As the Silurian drew to a close, the Taconic land barrier to the east finally disappeared to permit influx of fresh seawater from the east (Rickard, 1991a). During this time the carbonates of the Cobblskill, Glasco and Rondout formations were deposited.

Warm and clear epicontinental seas covered vast areas of the Appalachian basin in the early part of the Devonian when two prominent carbonate units, the Helderberg Group and Onondaga Formation, were deposited. The sea withdrew temporarily from all but southeastern New York after deposition of the Helderberg limestones resulting in progressive erosional truncation, from east to west, as far as central New York where the group is completely removed (Rickard, 1991b). A brief marine transgression occurred after the erosion of the Helderberg carbonates when the clastic sediments of the Tristate Group were deposited. Another short-lived marine regression allowed erosional truncation of the Tristate rocks. When the sea returned, deposition of the Onondaga Formation began under a vast stretch of epicontinental sea that rivaled the Helderberg sea in size. The Onondaga Formation is well known for its coral reefs in New York and Ontario.

2.2 The Acadian orogeny

The beginning of the tectonic disturbance known as the "Acadian orogeny" was signaled in the early Devonian by a widespread volcanic event when the Tioga ash fall, centered in Virginia, south of New York covered almost the entire basin (Faill, 1985). In New York the Tioga Metabentonite occurs within the Onondaga limestones at the base of its youngest member, the Seneca. Although the Acadian plate-tectonic event was centered, like the Taconic event, on New England and the Canadian Maritime Provinces, it was located slightly eastward and was quite distinct from its precursor.

The Acadian orogeny is believed to have resulted from a complex continent - continent collision between Laurentia, Armoria (Hercynian Europe), and possibly Gondwana with a number of terranes, or microcontinents, caught in between. On the basis of paleomagnetic data, Kent (1985) proposes that the "Traveler Terrane" (central New England and New Brunswick, Canada) rotated clockwise and was welded against Laurentia in the Devonian. This was followed by northward translation of the successively outboard terranes, Avalon and Meguma, and a counter-clockwise rotation of the Meguma. According to Williams and Hatcher (1982) and Ettensohn (1985a), however, the Acadian orogeny best coincides with the accretion of the Avalon Terrane (see fig. 2). In New Brunswick and Nova Scotia, evidence of deformation in the Upper Devonian and Carboniferous strata (Rodgers, 1967) suggests that there may not have been a significant break between the Acadian and succeeding Alleghenian orogeny in those areas.

Thus, unlike the Taconic event, which was marked by head-on collision of a continent and island arcs, the Acadian event was characterized by oblique convergence and transcurrent movement between continental blocks. Thus, large-scale overthursts, so typical of the Taconic orogeny, are unknown for the Acadian orogeny. Instead, the Acadian orogeny was marked by widespread magmatic intrusion, metamorphism, faulting and folding, mainly in New England and the Canadian Maritime Provinces (Rodgers, 1967, 1970; Price and Hatcher, 1983; Faill, 1985; Osberg, 1988). In New York, the best evidence for the Acadian orogeny is found in the deformation and metamorphism of the Precambrian rocks of the Hudson Highlands (Fakundiny et al. 1989).

Within the basin-proper, and even at its margin, evidence of Acadian deformation is scant and, at best, subtle. Arguments have been made for an Acadian age of the folded and faulted Silurian and Lower Devonian rocks of the Helderberg escarpment in eastern New York (see Marshak, 1986). These post-Taconian structures, however, have also been suspected to be of later Alleghanian origin (Zadins, 1984; Faill, 1985; Marshak, 1986). In general, it has been difficult to separate the subtle Acadian structures from the superimposed, stronger Alleghanian structures.

Within the basin, the strongest manifestation of the Acadian orogeny is neither structural nor magmatic, but sedimentological. An enormous volume of clastic sediments, stripped from the Acadian mountains, prograded over the northern and central Appalachians from northeastern source terrains during the Middle and Late Devonian and far exceeded the amount of sediment involved in the Taconic clastic wedge. The Acadian clastic wedge, popularly known as the "Catskill Delta", was a composite wedge and consisted of many pulses of alluvial fan-delta progradation (Rickard, 1991b, fig. 8.16).

In the Middle Devonian Hamilton Group, the occurrence of several thin limestone units (Cherry Valley, Stafford, Centerfield, Tichenor, Menteth and Portland Point members) in western and central New York probably indicates periods of interruption in supply of terrigenous sediments from the newly rising Acadian mountains (Kramers and Friedman, 1986; Savarese et al. 1986). Probably, the most significant break in the development of the Catskill deltaic sedimentation took place at the end of the Middle Devonian when the limestones of the Tully Formation were deposited over much of New York, with the exception of the eastern part (Johnson and Friedman, 1969; Heckel, 1973; Rickard, 1975).

The building of the "Catskill delta" resumed in full force from the beginning of the Upper Devonian; the landward facies prograded steadily over the seaward facies to the west throughout the Upper Devonian. This indicates that the subsidence of the basin was more than matched by sediment supply from the erosion of the Acadian Mountains during this interval. In fact, progradation of the subaerial part of the delta toward western New York was so rapid that the Catskill delta has been called a "tectonic fan-delta complex" (Friedman, 1988). No limestone units are known in the Upper Devonian of New York, and brief periods of relatively rapid deepening of the basin are probably indicated by tongues of several black shales confined to the western half of the state (Rickard, 1975). These black shales are missing from the latest Devonian Conneaut and Conewango groups indicating that toward the end of the Devonian the subaerial Catskill fan-delta complex had reached the western margin of New York State.

The sedimentologic record of the late Paleozoic is missing from most parts of New York. Scattered patches of the Lower Mississippian Knapp Formation and the Lower Pennsylvanian Olean Formation are found only along the southern border of New York. These are the northerly extensions of the post-Devonian clastic wedge (Pocono) of Pennsylvania. It is possible that these rocks once covered more extensive areas in New York but were subsequently removed by erosion (Friedman and Sanders, 1982; Johnsson, 1986; Gerlach, 1987).

The locus of Catskill deltaic deposition is believed to have shifted southward through the Devonian-Mississippian time in response to an oblique convergence with the Avalon microcontinent (Ettensohn, 1985a,b). The Lower Mississippian clastics of the Pocono Wedge in south-central Pennsylvania and southwestern Virginia and the equivalent shales (Chattanooga) of Tennessee and Alabama farther south probably represent the last stages of the Catskill deltaic sedimentation and the Middle Mississippian carbonates overlying these rocks mark the end of the Acadian Orogeny (Ettensohn, 1985a). However, if Aronson and Lewis (1994) are correct in their interpretation of K/Ar ages of detrital white mica in the Devonian - Middle Pennsylvanian rocks of eastern New York, western Maryland, and northeastern

Ohio, then the Acadian Orogen probably persisted as the major source for clastic sediment at least until the Middle Pennsylvanian time.

2.3 The Alleghanian orogeny

The late Paleozoic (Carboniferous - Permian) Alleghanian orogeny reorganized much of the Appalachian basin into its present structural configuration. The Alleghanian orogeny is distinct in that its most spectacular deformation, the folding of the Valley and Ridge Province south of New York, was within the basin, whereas the Acadian and Taconic deformations were confined to the eastern orogenic belt. The effects of the Alleghanian orogeny were felt most strongly in the southern and central Appalachian, whereas the Acadian and Taconic orogenies were centered in the northeastern edge of the basin. Also, unlike the previous two orogenies, the Alleghanian orogeny was marked by little magmatism.

A number of similarities, however, exist between the Alleghanian and the Taconic orogenies. Like the Taconic, the Alleghanian was a compressive tectonic event, this time associated with convergence and collision between Laurasia and Gondwanaland (Hatcher, 1978). Both orogenies were also marked by extensive thrusting. But, while Taconic thrusting was confined to the eastern basin margin, the Alleghanian thrusting was in the form of decollement in which the Paleozoic cover in the southern and central Appalachians moved northwestward along subhorizontal faults (decollements) above the basement into the basinal part (Gwinn, 1964; Perry, 1978; Hatcher and Zietz, 1980; Price and Hatcher, 1983; Faill, 1985). The net northwestward displacement of the detached rocks is estimated at 200km in Alabama, Tennessee and Kentucky and about 100km in Pennsylvania (Gwinn, 1964; Cook et al. 1979; Ando et al. 1983). Fault splays originating from the master decollement zone created folds, step faults and listric thrust faults in the Valley and Ridge and the Plateau provinces (Gwinn, 1964; Faill, 1985). These rootless structures gradually become diminutive from southeast to northwest into the Appalachian Plateau of northwestern Pennsylvania and south-central New York, where only broad and open folds are found.

In New York, deformation that may be attributed to the Alleghanian orogeny includes (1) the broad and open folds in the plateau in south-central New York, north of the Pennsylvania border, (2) tight folds and thrust faults in a "miniature Valley and Ridge Province" along the Hudson Valley, eastern New York (Marshak, 1986), and (3) mesoscopic and microscopic structures like joints, mechanical twins, solution cleavage, crenulation cleavage, pencils and deformed fossils (Engelder and Engelder, 1977; Engelder and Geiser, 1979, 1980; Geiser and Engelder, 1983).

The open folds along the southern border of New York have limbs that dip at angles of only 1-2 degrees and trend roughly ENE-SWS (Tillman and Barnes, 1982). These folds may have resulted from decollement tectonism located in the subsurface Salina rocks (Gwinn, 1964; Jacoby and Dellwig, 1974) and mark the northern termination of large-scale Allleghanian structures in the Appalachian Plateau.

The fold-thrust belt of the Hudson Valley in eastern New York has much smaller-scale structures than those in the Valley and Ridge province of Pennsylvania with which it is apparently continuous. The trend of the structures in the Hudson Valley is also different, NNE-SSW, as opposed to the NE-SW trend in the Valley and Ridge. There is considerable disagreement as to the age of these structures in the Hudson Valley: both Acadian and Alleghanian have been proposed (see Marshak, 1986 for discussion).

The meso- and microscopic structures found in the Paleozoic rocks, primarily Devonian, of New York are believed to represent "layer-parallel shortening fabrics" resulting from the Alleghanian orogeny (Engelder and Geiser, 1979, 1980; Geiser and Engelder, 1983). These authors recognize two non-coaxial phases of Alleghanian deformation from these fabrics. The effects of the early or "Lackawanna Phase" deformation are found mainly in the Hudson Valley and are interpreted as a possible product of strike-slip motion between the Avalon microcontinent and North America. The shortening fabrics that occur throughout the Plateau as well as the Valley and Ridge Province, are parallel to the trend of major structures, and have been attributed to the "Main Phase" of the Alleghanian orogeny, which has been correlated with the final convergence of Africa against North America and the accreted terranes.

It has proved difficult to firmly establish the timing of the Alleghanian orogeny. Structural evidence from the southern Maritime Province suggests a nearly continuous or spasmodic orogenic movement from Middle Devonian to early Permian with no clear boundary separating the Acadian from the Alleghanian events (Rodgers, 1967). In the Appalachian Plateau and Valley and Ridge Province, the Alleghanian movement seems to have begun in the post-Lower Permian (Rodgers, 1967; Van der Voo, 1979), while in the southern Appalachian the movement may have begun as early as the Mississippian (Rodgers, 1967). According to Geiser and Engelder (1983) the "Lackawanna Phase" of Alleghanian deformation in New York was probably initiated in the Pennsylvanian, whereas the "Main Phase" began in the early Permian.

2.4 Post-Alleghanian period

There is no direct sedimentological record of the end Devonian - Late Triassic interval in New York, although much has been written about the missing post-Devonian strata (see "Previous Studies"). Triassic - Jurassic rocks of the Newark Supergroup are preserved only in Rockland County of southeastern New York (fig. 3). These rocks represent the northeastern extremity of the Newark-Gettysburg Basin, the largest of the onshore Triassic basins of eastern North America.

These basins began forming in the Late Triassic as the late Paleozoic supercontinent drifted north, possibly over a hotspot (Morgan, 1980) or across a tensional stress field (Bedard, 1985). The crust was pulled apart along old fractures, sutures and transforms, created during the three preceding plate tectonic events, and rifting of the crust and the definitive opening of the Atlantic Ocean began. Extension along proto-Atlantic axes led to the formation of at least 30 clastic and evaporitic 'synrift' basins near the margins of the North American and African plates (Manspeizer and Cousminer, 1988). Onshore in eastern North America, these basins are now part of the Piedmont Province. Recurrent subsidence due to downwarping of the ductile crust followed by faulting in the brittle crust allowed accumulation of huge quantities of terrestrial sediments in these small basins amounting to 7-9km in the Newark-Gettysburg Basin alone (Sanders, 1963; Olsen et al. 1982).

The Triassic basins were closed basins with internal drainage and are characterized by fluvial sandstones and conglomerates that thicken toward the axis of the basins where they interfinger with deep water lake deposits (Manspeizer and Cousminer, 1988).

In New York, the rocks of the Newark Supergroup (Stockton and Brunswick formations) were intruded by the Palisade diabase sills at about 201 Ma in the early Jurassic time (Dunning and Hodych, 1990). Similar sills and flood basalts of early Jurassic time occur in nearly all the Triassic basins of eastern North America (Boer et al. 1988).

Another Mesozoic magmatic event in New York, unrelated to the Triassic basins, is represented by kimberlite dike swarms of the eastern Finger Lake district (fig. 3). Most of these kimberlites crop out as vertical dikes in prominent N-S joint sets (Kay et al. 1983). Radiometric dates (~140 Ma) suggest a Late Jurssic to Early Cretaceous age for some of these dikes (Basu et al. 1984). These intrusions, along with other kimberlites, occur along a trend extending from Tennessee to New York coinciding with the "Keel Line", or the deepest part of the Appalachian Basin (Dennison, 1983; Kay et al. 1990). According to Dennison (1983), the dikes were intruded along fractures that developed along the Keel Line as a result of stretching due to post-Alleghanian isostatic rebound. Mantle fragments in some kimberlites suggest that the magma was derived from the mantle at depths of 150km or more (Kay, 1990).

Evidence for mixed Mesozoic-age uplift in New York were cited in Tillman and Barnes' (1983) work with fluid inclusions in mineralized faults of west-central New York and fission-track study of Miller and Duddy (1986), Duddy et al. (1987), and Miller (1990).

The post-Cretaceous was a time of extensive erosion when the modern landscapes of New York and much of the Appalachians were sculpted (Rodgers, 1967). Much of the eroded sediments has been transported by rivers and more recently by glaciers on to the Atlantic continental shelf, some of which has been deposited in the Coastal Plain Province. In New York, Cretaceous and younger deposits (Raritan and Magothy formations) of the Coastal Plain are confined to parts of Long Island and Staten Island (fig. 3). Evidence of continuing uplift in the Adirondacks has been cited by Isachsen (1985, 1992).

The lack of sedimentological record requires that the entire post-Devonian geological reconstruction of New York be based on several lines of indirect evidence, which will be discussed in the following chapters. It would seem likely that erosion dominated the post-Devonian geologic history of New York, but how much erosion took place and when it began remain to be resolved.

3 PREVIOUS STUDIES IMPLYING POST-DEVONIAN SEDIMENTATION IN NEW YORK

Prior to the 1980s, it was generally believed that since the end of Devonian time little sediment had accumulated in the northern Appalachian Basin of New York, and the Alleghanian Orogeny had failed to have much impact on sedimentation in this area (Meckel, 1970). This interpretation was primarily based on the absence of post- Devonian strata from much of New York and apparent thinning of post-Catskill strata to the north from Pennsylvania (Wood et al. 1969; Edmunds, 1979).

Over the last fifteen years, however, a significant number of publications have challenged the former notion of little or no post- Devonian sedimentation in New York. Most of these studies, discussed below, are based on various paleotemperature signatures of the surficial or shallow subsurface rocks in New York and adjoining areas. The inferred paleotemperatures are rather high and, generally have been interpreted as being acquired during deep burial. Deep burial of these rocks requires the former presence of post-Devonian strata much thicker than previously thought.

The earliest works that allude to former deep burial of the surface-exposed strata in New York are perhaps the Conodont Alteration Index (CAI) maps of the Appalachian basin presented in Epstein et al. (1977), Harris et al. (1978), and Harris (1979). As seen in figure 5, their CAI values for Silurian-through-Middle Devonian carbonates range from 2 to 4.5 west to southeast along the northern erosional edge of these rocks in New York State. If the Middle Devonian carbonates of New York (approximately 380 m.y old) began uplifting in the late Pennsylvanian, coinciding with the onset of the Alleghanian orogeny, then the maximum time of burial and heating was about 100 m.y (see Harris, 1979). Using Harris' (1979, fig. 2) method of calculating the lowest possible "maximum paleotemperature" from CAI and burial time, a temperature range of 110 - 250° C is obtained for the CAI range of 2 - 4.5. If uplift began in the Middle Triassic, coinciding with the rifting of eastern North America, a maximum burial time of 180 m.y can be assumed (Harris, 1979), and paleotemperatures of 100 - 240° C are obtained for the same CAI range. If a paleogeothermal gradient of 30°C/km and a mean annual surface temperature of 20°C (see section 7.1) are used, these temperatures translate into about 2.5km post-Middle Devonian overburden in western New York and a staggering 7.6km along the eastern border of New York. If one subtracts a projected thickness of 1.5km Devonian strata younger than Middle Devonian carbonates (i.e., the Tully Limestones and equivalents) from western New York and 3km from eastern New York (Rickard, 1975), a former presence of 1.5km post-Devonian strata in western New York and 4.5km in eastern New York is obtained.

Conodont study by Legall et al. (1981) in Ontario and Quebec provinces of Canada also shows CAI isograds of 3 to 4.5 in Ordovician carbonates just west and north of New York which are generally consistent with Epstein et al.'s (1977) isograds farther south and may imply the former presence of post-Devonian sediments over these areas of Canada as well, although Legall et al. (1981) proposes anomalous geothermal heat as the cause of high conodont alteration values.

On the basis of conodont data from Epstein et al. (1977), the presence of Paleozoic outliers in the Canadian Shield, and isopach trends of preserved Paleozoic rocks in the Northern Appalachian basin, Crough (1981) concluded that southeastern Canada and New England were elevated and eroded at least 4km relative to the Central Appalachian basin. Citing fission track dates, Crough (1981) suggested that the uplift took place in the Cretaceous - early Tertiary time when the northeastern part of North America moved over a hotspot.

Friedman and Sanders (1982), on the basis of vitrinite reflectance of anthracite-grade plant debris, alteration index of associated carbonized kerogen,

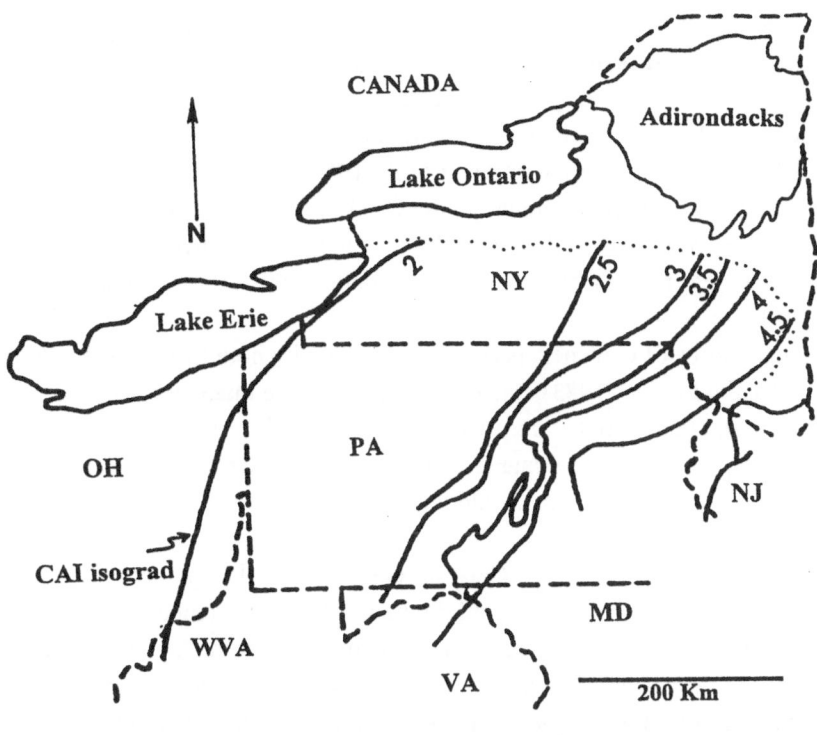

················· Erosional edge of Middle Devonian strata

Figure 5: Conodont color alteration index (CAI) isograd map for Silurian through Middle Devonian limestones in central and northern Appalachian basin (modified from Harris, 1979). Abbreviations: NY = New York, NJ = New Jersey, PA = Pennsylvania, OH = Ohio, WVA = West Virginia, VA = Virginia, MD = Maryland.

conodont alteration index, and authigenic mineralogy in the uppermost Middle Devonian Gilboa Formation of the eastern Catskill Mountains, concluded that the Catskill Mountains were once buried to a depth of about 6.5km. They used Hood et al.'s (1975) method to calculate the maximum paleotemperature (190° C) from a vitrinite reflectance of 2.5% and an assumed heating time of 200 m.y. Then using a paleogeothermal gradient of 26°C/km and a mean annual surface temperature of 20°C, they calculated a burial depth of 6.5km for the Gilboa samples.

Friedman and Sanders' (1982) study came under attack from Levine (1983), who questioned the various assumptions and measurement errors inherent in Hood et al.'s (1975) method, the choice of a long heating time and a low geothermal gradient and the use of vitrinite reflectance data from a single sampling site. Friedman and Sanders (1983) responded that while the use of higher geothermal gradients makes a considerable difference in their estimate of burial depths, a much lower heating time (35 m.y) increases the burial depths by only about 1km for any geothermal gradient used. Friedman and Sanders (1983), however, show that a vitrinite reflectance gradient of 1.4%/km measured in boreholes of the Southern Anthracite Field, Pennsylvania by Levine (1983) was too high for the Catskills. For a vitrinite reflectance of 2.5% in the Gilboa samples, this gradient means a burial depth of only 1.79km, but a paleo- geothermal gradient of 91-105° C/km. No evidence exists for such a high paleogeothermal gradient in the Catskill region. The basal Paleozoic rocks in the area, 4.5 km below Gilboa, show no evidence of metamorphism which would be expected if such a high geothermal gradient existed (Friedman and Sanders, 1983).

A different source of evidence, perhaps overlooked by both Friedman and Sanders (1982, 1983) and Levine (1983), and one that makes Friedman and Sander's estimate of burial depth not too surprising, is stratigraphic. At least 2.4 km of post-Gilboa, Upper Devonian strata are preserved in western New York (Rickard, 1975) and depositional thickness would have increased towards the Catskill region and the Acadian sediment source to the east. Having projected a minimum thickness of 2.4 of Upper Devonian over the Catskill region, about 4km (not 6.5km) post-Devonian cover is required to explain the vitrinite reflectance of 2.5% using Friedman and Sanders' (1982) method. This estimate is not that much different from the thickness of post-Devonian strata (~ 4.5 km) over eastern New York implied by Epstein et al.'s (1977) CAI maps. If one uses a higher geothermal gradient (for example 30° C/km, approximately the present average geothermal gradient of New York, Hodge et al. 1982) and the mean vitrinite reflectance (not the maximum), the estimated thickness of post-Devonian strata in this area decreases to a little less than 3km.

Tillman and Barnes (1983), on the basis of fluid-inclusion study of mineralized faults and fractures in the Upper Ordovician Oswego sandstones of northwestern New York, inferred that fracturing and mineralization represented two episodes of post-Alleghanian deformation at temperatures between 176° C and 120° C and 112° C and 76° C. A paleogeothermal gradient of 25° C/km, used in other burial studies in New York (see Friedman and Sanders, 1982; Lakatos and Miller, 1983), would mean a former maximum burial depth of about 6km for these rocks, but Tillman and Barnes (1983) rejected it as "too high a figure" and inferred that the paleogeothermal gradient of the study area was locally elevated as a result of migration of warm basinal fluid originating from clay dehydration at a deeper level.

Lakatos and Miller (1983), in their fission-track study of apatite and zircon particles, concluded that the lowermost Upper Devonian strata in the Catskill area were buried to about 4km in the last 125 m.y and between 4 and 7km prior to that period. The conclusions of their study concur with those of Friedman and Sanders (1982).

Urschel and Friedman (1984), on the basis of fluid inclusion, oxygen isotope, and vitrinite reflectance studies of the Lower Ordovician Beekmantown carbonates of the upper Hudson and Champlain valleys concluded that these rocks may have been buried to depths of as much as 7km. If one assumes that the 5km thick Middle Ordovician to Upper Devonian section elsewhere in New York (Rickard, 1975; Fisher, 1977) once extended to this area, a 2km post-Devonian overburden is necessary to account for the 7km burial.

Discovery of fragments of Middle Devonian sedimentary rocks containing a distinct Hamilton fauna in a Cretaceous age diatreme near Montreal, Canada indicates the presence of Devonian strata north of New York State at least until the Cretaceous (Clark, 1972; Boucot et al. 1986).

Gale and Siever (1986) studied the diagenesis of Catskill facies sandstones in southeastern New York. They concluded that these rocks "suffered deep burial during the late Paleozoic and uplift and erosion after the Allegheny Orogeny".

On the basis of vitrinite reflectance, sediment compaction and fluid-homogenization temperature study in the anthracite district of northeastern Pennsylvania, Levine (1986) inferred a former burial depth of 6-9km for the extant Carboniferous strata. However, according to Levine, the paleomagnetic ages of folding and paleobotanical ages of the youngest rocks in the area delimit the maximum time of burial and deformation to a relatively short interval (290 - 270Ma). Since high sedimentation rate required for the proposed burial of these rocks in such a short time would be geologically unrealistic, he proposed rapid tectonic burial under Alleghanian overthrusts as a possible burial mechanism. Although structural evidence for a large overthrust in the area is lacking, Levine (1986) contends that thrusting could be in the form of smaller imbricate slices which could at least serve as proximal sources for very rapid sedimentation. One could argue that a similar mechanism of burial could have operated in southeastern New York, not far from the northeastern anthracite region and cause the deep burial of the Devonian strata (Friedman and Sanders, 1982; Lakatos and Miller, 1983). Levine's (1986) study thus shows a significant departure from his earlier contention (Levine, 1983) that the Catskill area could not have been buried under thick Carboniferous overburden. However, Daniels et al. (1990), in their study of authigenic clay minerals in inferred hydrothermal veins in coal of eastern Pennsylvania, proposed that the late Paleozoic geothermal gradient in the region was high and anthracitization probably took place at depths of 5km or less (as opposed to 6 - 9 km suggested Levine, 1986).

On the basis of fission-track and clay-diagenesis study of the lower Middle Devonian Tioga bentonite (within Onondaga limestones) across New York State, Johnsson (1986) suggested that this geologically isochronous unit might have been buried to about 7km in eastern New York. Projecting the thickness of the post-Tioga Devonian section of New York onto the outcrop belt of the Tioga Bentonite, he inferred that nearly 4km of post-Devonian sedimentary rocks once covered the area (fig. 6). This estimate is consistent with those made in Epstein et al. (1977), Crough (1981), Friedman and Sanders (1982), and Lakatos and Miller (1983).

Karig (1987) questioned Johnsson's (1986) interpretation of deep burial of the Tioga Bentonite bed on the ground that the low uniform paleogeothermal gradient of 20 - 25° C/km used by Johnsson to calculate the burial depths may have been inappropriate. For example, the abrupt increase in inferred paleotemperature (i.e, low fission-track age and high illite concentration) in central New York, could have resulted from a thermal anomaly associated with the kimberlite dike swarms of the eastern Finger Lake area (fig. 3). Johnsson (1987) replied that the occurrence of these dikes also coincided with the axis of greatest sediment accumulation, or the "keel line", in the Appalachian basin (Dennison, 1983), so that it was difficult to separate out the paleotemperature signatures of rocks of central New York originating from deep burial from those imprinted by the dikes.

In Cherry Valley of east-central New York, Gurney and Friedman (1987) studied fluid inclusions, oxygen isotopes and vitrinite reflectance of the Lower Devonian Helderberg and lower Middle Devonian Onondaga limestones. They inferred a former burial depth of about 5km for these rocks. Projection of the 2.5km thick post-Tioga Devonian section of southern New York onto Cherry Valley indicates a former presence of about 2.5km post-Devonian strata. This estimate is consistent with the result of Johnsson's (1986) work, which included Tioga samples from the same Cherry Valley outcrops, but were based on entirely different techniques.

Similarly, fluid-inclusion studies of the Middle Silurian Lockport dolostones of western New York suggest a former burial depth of 5km (Friedman, 1987a, 1987b). A lower estimate of 4km burial is given by Gross and Engelder (1989), who studied the fracture system and fluid-homogenization temperatures of the fracture-filling calcite in the same rocks. A former presence of 2.5 to 3km post-Devonian strata can be inferred for the belt of Lockport outcrops in western New York, if the thickness of the post-Lockport rocks between southwestern and southcentral New York is projected onto the belt.

Gerlach's (1987) Lopatin modeling, based on a regional vitrinite reflectance trend in the Upper Devonian Rhinestreet Formation (black shale) in south-central New York, shows that a northward extension of the Carboniferous strata from Pennsylvania is necessary to account for the high thermal maturities observed in the studied rocks. His estimate of the "minimum" thickness of the Carboniferous rocks in the study area ranges between 1.20 and 3.37km.

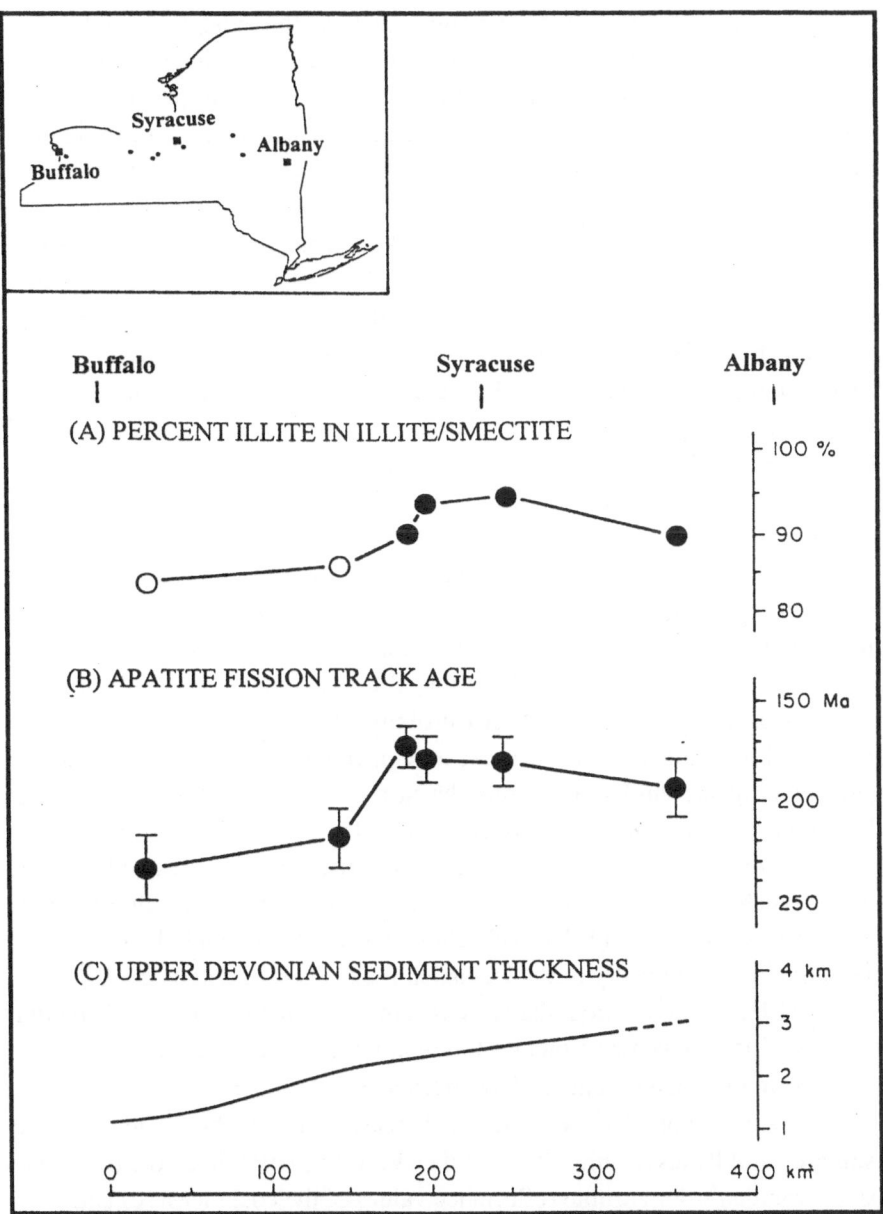

Figure 6: West to east profiles of (A) percentage of illite in illite/smectite mixed-layer clays and (B) apatite fission-track ages of samples of Middle Devonian Tioga bentonite across New York, and (C) estimated thicknesses of overlying Upper Devonian strata projected on the sampling locations from southern New York. (From Johnsson, 1986).

Beaumont et al. (1987) used a mathematical- geodynamic model to simulate the possible effects of the Alleghanian orogeny on the Late Paleozoic evolution of the sedimentary basins of eastern interior North America. They reconstructed the maximum sediment thickness distribution in the Appalachian basin at the height of the Alleghanian orogeny by using moisture-content data from near-surface coals. The model that agrees with these data predicts that the Alleghanian overthrusts had thickened the crust by as much as 20km in the Appalachian and Ouachita orogens. The predicted erosion required to bring the coals to the surface exceeds 15km in some areas. In New York, according to their model, post- Alleghanian erosion ranged from about 3km in western New York to about 9km in southeastern New York (fig. 7). It is interesting to note the similarity of this predicted value for southeastern New York with the 6-9km burial of the Carboniferous strata in northeastern Pennsylvania inferred by Levine (1986).

Jackson et al. (1988) studied the authigenic magnetite content of the Helderberg (Lower Devonian) and Onondaga (lower Middle Devonian) carbonates along the east-west outcrop belt extending from Albany, New York to southeastern Ontario, Canada. Like the fission-track age and illite-content profiles of Johnsson (1986), the magnetite-content profile shows a peak in central New York. Previous paleomagnetic studies indicate that the authigenic magnetites had formed no later than late Paleozoic. Jackson et al. (1988) attributed the correlation between high magnetite and illite content of these rocks to temperature-dependent diagenesis triggered by orogenic fluid derived from the east: the maximum burial temperatures and hence the strongest diagenetic effects occurred in central New York.

Miller and Duddy (1989) studied apatite fission-track ages of surficial Middle to Upper Devonian sandstones across New York State and obtained results similar to those of Lakatos and Miller (1983) and Johnsson (1986). Although their interpretation of uplift history of the Devonian rocks differ in details from Johnsson's (1986) interpretation, they also reached the conclusion that substantial thicknesses of post-Devonian strata - ~2-3km in western New York and "greater than 3-4km" in the Catskill region - have been lost.

In a recent study of the Pennsylvanian coal measures in nine boreholes of western and south-central Pennsylvania, Zhang and Davis (1993) calculated the thicknesses of lost overburden from vitrinite reflectance values of the coal measures and constructed a Permian isopach map for the area (fig. 8). The isopachs of lost Permian strata (Zhang and Davis, 1993) trend NE-SW increasing in thickness from the northwest to southeast, much like the Devonian strata (Faill, 1985), and suggest that Permian strata - as much as 3.5 km in thickness - might once have extended into New York State. Works of Paxton (1983), Orkan and Voight (1985) and Lacazette (1991) support Zhang and Davis' (1993) conclusions.

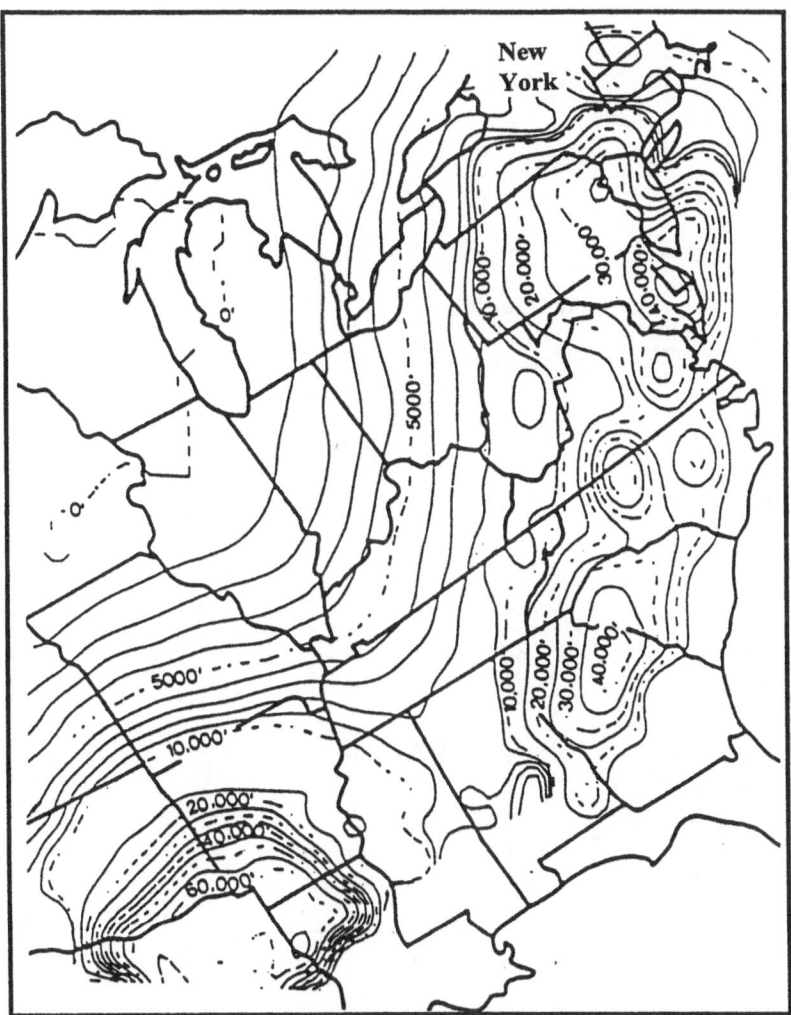

Figure 7: Isopachs (in feet) of predicted missing strata from the Appalachian basin due to post-Alleghanian erosion, based on mathematical geodynamic model of Beaumont et al. (1987).

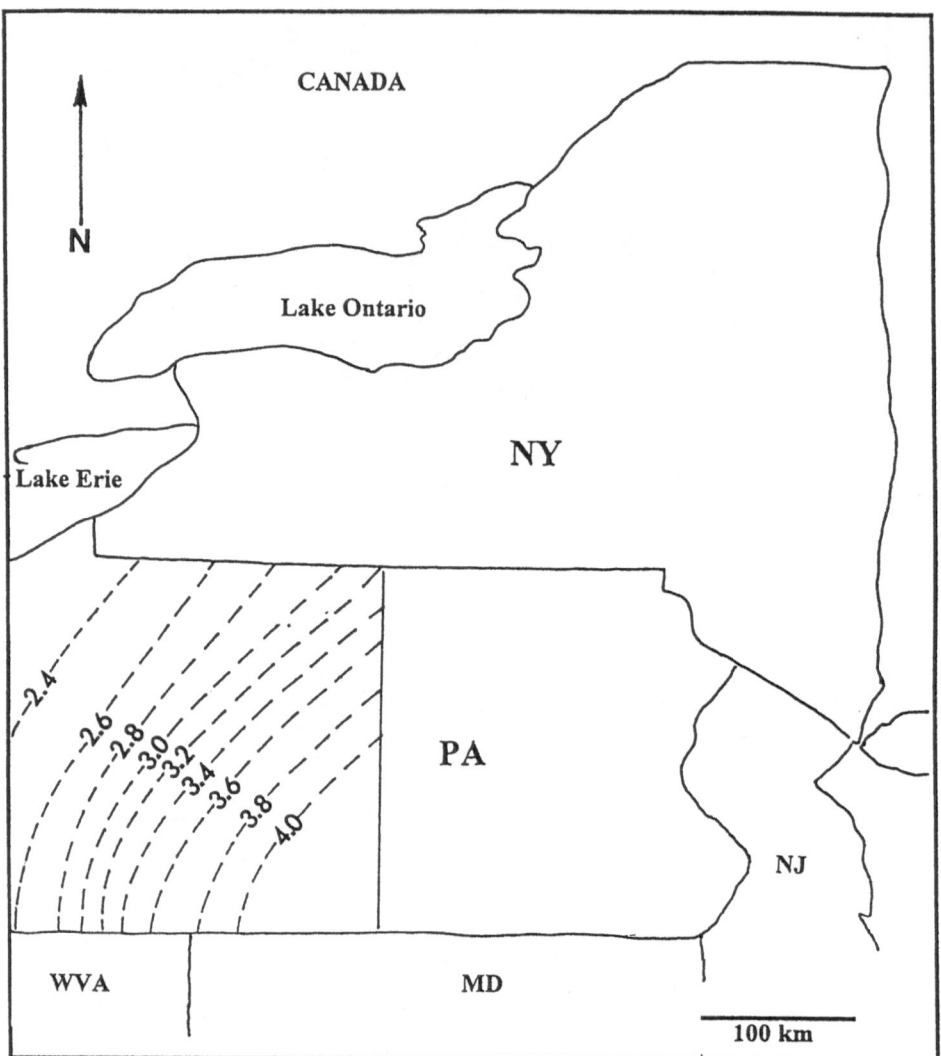

Figure 8: Map showing isopachs of inferred Permian strata in western Pennsylvania drawn by Zhang and Davis (1993) on the basis of vitrinite reflectance study of the Pennsylvanian coal measures. The NE-SW trend of the isopachs indicates the former presence of the same strata in New York. Abbreviations: NY = New York, PA = Pennsylvania, WVA = West Virginia, NJ = New Jersey. (Modified from Zhang and Davis, 1993).

Most of the studies discussed above deal with small areas and single rock units. With the exception of the studies by Johnsson (1986), Jackson et al. (1988) and Miller and Duddy (1989), spatial variation in paleotemperature and burial estimates are not shown even in studies that cover relatively large areas. To reconstruct a comprehensive scenario of the burial history of the Paleozoic rocks in New York, the spatial variations in various paleotemperature signatures of selected rock units have to be determined.

4 RESEARCH METHODS

4.1. Sampling

Sampling for the proposed study had to meet two important requirements: (1) it had to be representative of the Paleozoic sequence over the study area as much as possible, and (2) the samples needed to be suitable for laboratory analyses aimed at determining paleotemperatures.

Carbonates with grain-rich fabric (grainstones and packstones) were selectively sampled as these provide the best cements for fluid inclusion and isotope analyses. However, cement-filled vugs and fractures in any rock type were also collected for the same analyses. For organic maturation analysis, black shales were primarily used, but some samples of dark, organic- rich carbonates were also collected and analyzed. Any shale could have been used for clay-diagenesis study, but black shales were chosen because the results supplemented organic maturation data and thus served as better indicators of paleotemperature. The names and ages of the rock units sampled for this study are given below.

Upper Devonian: Dunkirk, Geneseo, Rhinestreet, and Middlesex black shales.

Middle Devonian: Tully Limestone
                  Marcellus Formation (black shale)
                  Onondaga Formation (limestone)

Lower Devonian: Helderberg Group (limestone)

Upper Silurian: Salina Group (carbonates, evaporites, and shale)

Middle Silurian: Lockport Group (dolostones and limestones)

Lower Silurian: Clinton Group (carbonate components only)
       Medina Group (shale only)

Upper Ordovician: Utica Formation (black shales),
       Trenton Group (limestones),
       Black River Group (limestones)

Upper Cambrian-
Lower Ordovician: Beekmantown Group (dolostones and sandstones)

Samples were collected from seventy-three locations in New York state and two locations in the Ontario Province of Canada (fig. 9). Fifteen of these are core sample locations. The core samples were provided by New York State Geological Survey, and include samples drilled by the Department of Transportation (DOT) of New York, the Supercollider Study Project, the Eastern Gas Shale Project (EGSP), and various other mining companies. The rest of the samples were collected from outcrops.

4.2 Petrography

Standard thin sections of carbonates and some shale samples with mineral veins were prepared and examined in order to identify types of cements and establish a paragenetic sequence among the various diagenetic components.

Late-stage cements post-date early cements and most compaction features, and may post-date stylolites and fractures. Such cements often include clear, poikilotopic calcite (Moore, 1984), xenotopic dolomite crystals (Friedman, 1965; Gregg and Sibley, 1984), and saddle dolomite (Radke and Mathis, 1980). Identification of late-stage cements is important, because they generally form in the burial environment, and fluid inclusion and isotope data from these cements may reveal the paleotemperatures and extent of burial experienced by the host rocks. Staining techniques (Friedman, 1959) have also been used for quick identification of calcite and dolomite.

4.3 Fluid inclusions

Fluid inclusions are small, usually microscopic, samples of fluid trapped within a mineral crystal (figs. 10, 11). They may be trapped in various growth irregularities

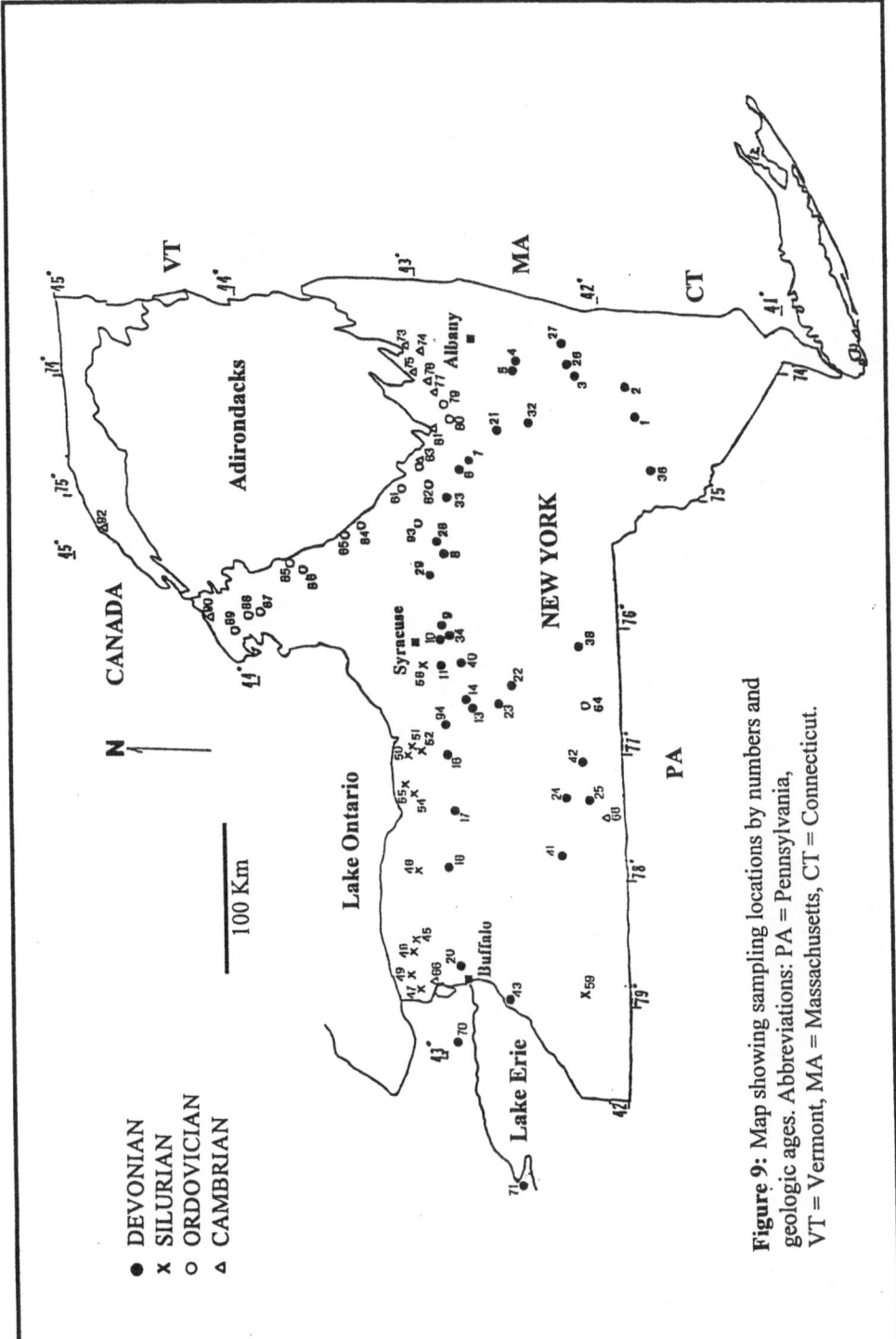

Figure 9: Map showing sampling locations by numbers and geologic ages. Abbreviations: PA = Pennsylvania, VT = Vermont, MA = Massachusetts, CT = Connecticut.

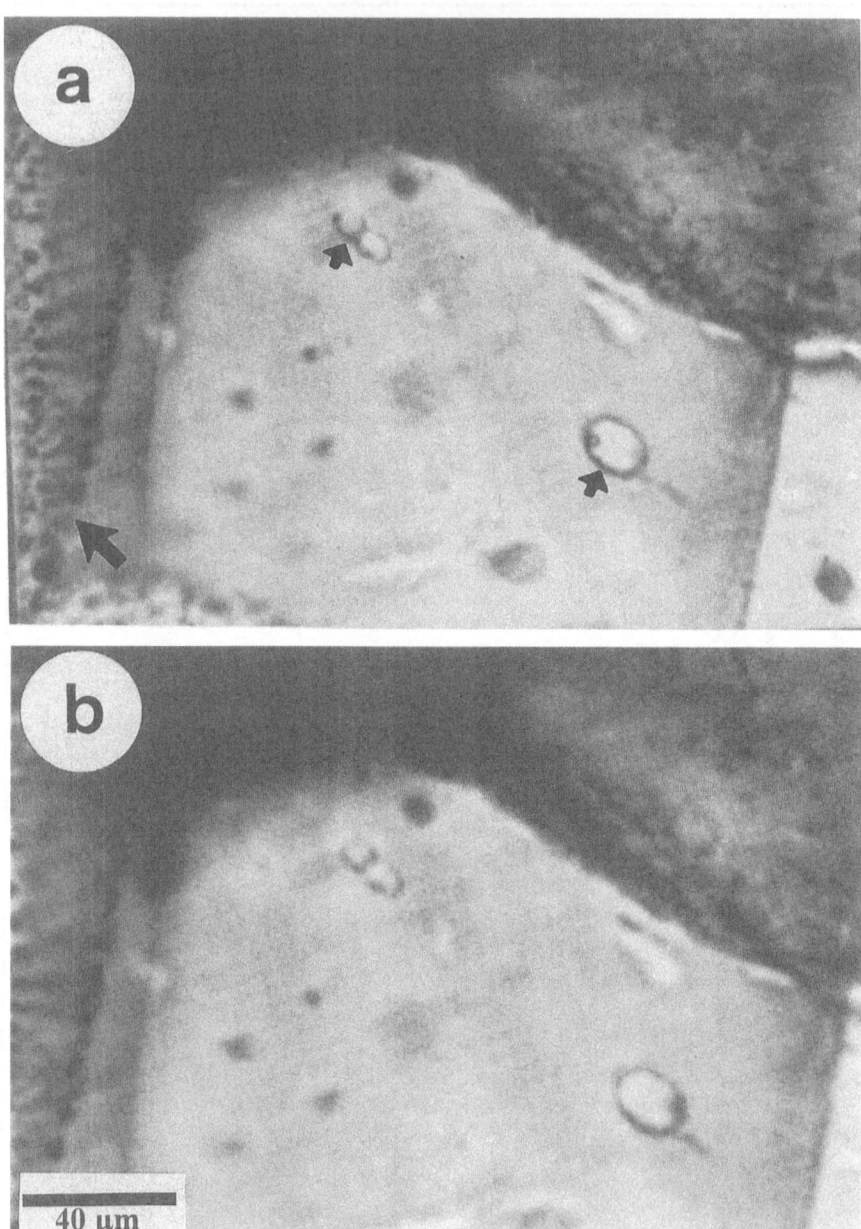

Figure 10: Photomicrographs showing two-phase (liquid + vapor) primary inclusions (small arrows) in fluorite (a) before heating and (b) after homogenization into the liquid phase. Lines of secondary inclusions (thick arrows) cut across crystal boundaries. (Sample from Helderberg limestone, location 9)

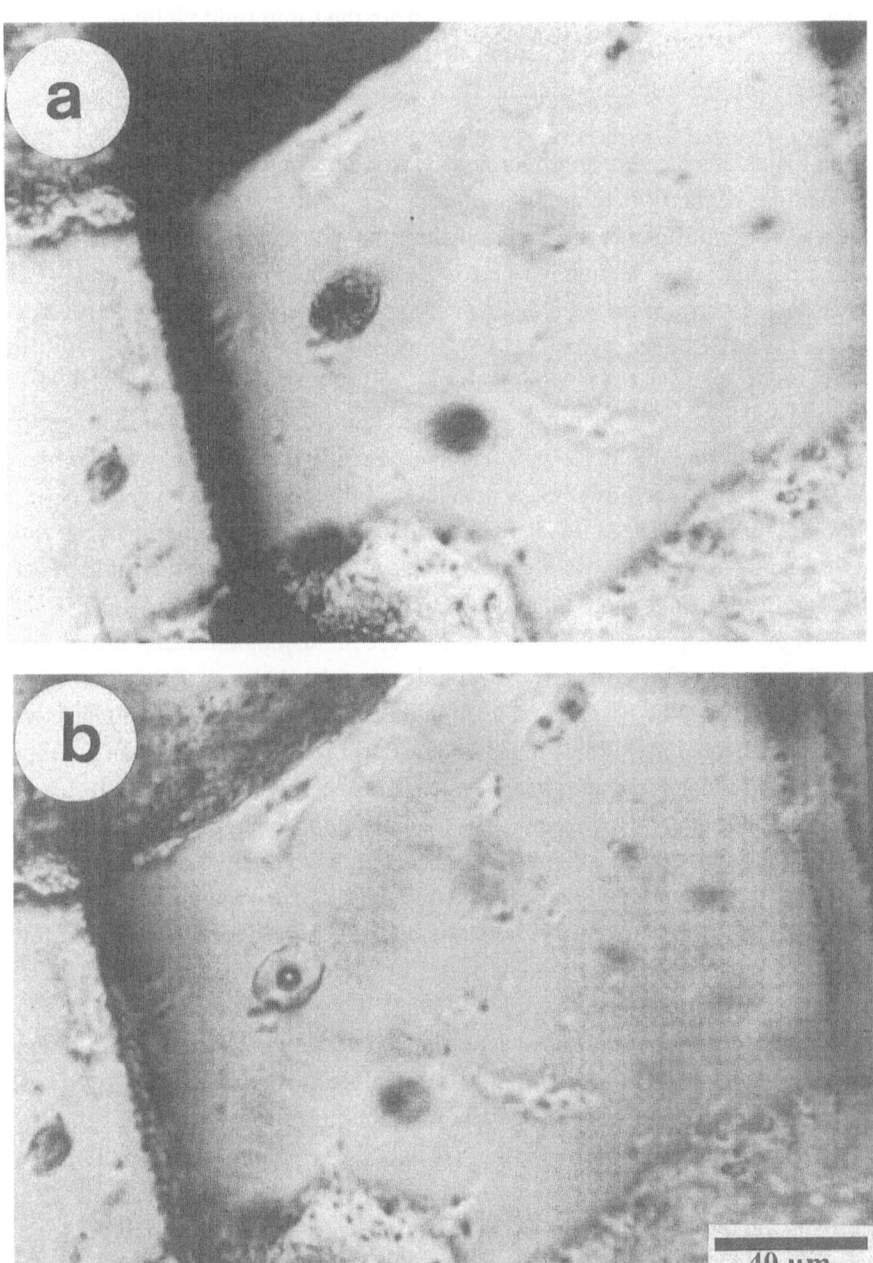

Figure 11: Photomicrographs of the same inclusions in figure 10 showing (a) their frozen state attained through supercooling and (b) return of the vapor bubble after the last ice crystals have melted as a result of slow heating.

or 'defects' in the crystal during its growth from the same fluid yielding "primary" fluid inclusions, and in microfractures that were healed during crystal growth or at a later time yielding "pseudo-secondary" and "secondary" fluid inclusions, respectively.

Fluid inclusions provide an important tool for determining paleotemperatures and compositions of the fluid at the time of trapping and have been widely used in studies of ore deposits, igneous and metamorphic petrogenesis, sediment diagenesis, and thermal history of sedimentary basins. Fluid inclusion techniques and applications have been covered extensively by Hollister and Crawford (1981), Roedder (1984), Shepherd et al. (1985), Burruss (1988), and recently by Goldstein and Reynolds (1994) and De Vivo and Frezzotti (1994).

Most inclusions are trapped as a one-phase fluid (Roedder and Bodnar, 1980; Burrusss, 1987) and the inclusions may subsequently change into two or more phases. Most common, however, is the change into a liquid and a vapor phase. According to Sorby (1858) this happens due to differential shrinkage of the fluid and the enclosing crystal possibly in response to subsequent cooling of the host rock. The fluid shrinks more than the crystal resulting in a vacuum which is occupied by the vapor phase. In such two-phase inclusions the vapor phase generally takes the form of a circular bubble (figs. 10, 11).

In the laboratory, the two phases in an inclusion change back to one phase when the crystal is heated in a Fluid-Inclusion Heating-Freezing Stage. Upon heating, the liquid in the inclusion expands at the expense of the vapor bubble which steadily shrinks and ultimately disappears (fig. 10). The opposite happens when the vapor/liquid volume ratio is large. The temperature at which one of the phases disappears, that is, one homogeneous phase appears, is called the "homogenization temperature" (T_h). Homogenization temperatures generally provide a minimum estimate of the trapping temperature of the inclusion, but with 'pressure correction' the trapping temperature can be approximately determined (see below).

Fluid inclusions are also used to determine composition of the fluid from which the crystals have precipitated, although the smallness of most inclusions pose severe technical problems. The simplest of all the techniques for determining fluid composition is the study of the freezing behavior of the inclusions. In this method, a given inclusion is frozen in a standard Heating-Freezing Stage by passing depressurized cold nitrogen gas over the sample. Because of the small size of inclusions, generally considerable supercooling (sometimes below -100° C) is necessary to freeze the fluid into ice crystals (fig. 11). In very small inclusions, even the ice crystals are generally not observed, and when the vapor bubble suddenly shrinks or disappears, freezing is supposed to have occurred. This freezing temperature is, however, not the true freezing temperature of the inclusion fluid. To determine true freezing temperature, the frozen inclusion is gradually reheated and the temperature at which the last ice crystal melts (fig. 11b) is considered the true freezing temperature (also called 'ice-melting temperature', or T_m). In very small inclusions T_m is measured at the point when the vapor bubble suddenly reappears, or makes a final sweeping movement toward its pre-freezing position in the inclusion.

Use of T_m as composition indicator is based on the assumption that the inclusion contains a simple NaCl-H$_2$O solution. From experimental studies, the dependance of freezing-point depression of water on salinity (wt. % NaCl) is well known. The salinity of inclusion fluids can be determined from the following relationship (Potter et al. 1978):

$$Ws = 1.769580 - 4.2384 \times 10^{-2}O^2 + 5.2778 \times 10^{-4}O^3 \pm 0.028,$$
where Ws = the weight percent NaCl in solution and
O = freezing point depression in °C.

Inclusion fluids may, of course, contain other salts, but it is generally agreed that for mixed Ca-Na-K-Mg chloride solutions, the error in estimating salinity (NaCl-equivalent) using the 'freezing-point depression' method is less than 5% (Clynne and Potter, 1977).

Another method of qualitatively determining the presence of the dominant salt is from the 'temperature of first melting (T_{fm})' of the ice crystals. T_{fm} is believed to represent the eutectic temperature. Where there is no solid-solution between the solid phases, eutectics are invariant points and, therefore, the first melting occurs at a temperature determined solely by the composition of the end-member components. For example the eutectic temperature of NaCl-H$_2$O system is -20.8°C, CaCl$_2$-H$_2$O is -49.5°C, KCl-H$_2$O is -10.6°C and CaCl$_2$-NaCl-H$_2$O is -52°C (Shepherd et al. 1985). This potentially important method is, however, seriously constrained by the difficulty of identifying the first melting of ice, especially in small inclusions typical of sedimentary rocks.

The problems and limitations of fluid-inclusion studies have been addressed by several authors (see the above references). Succinct treatment of the subject will be found in Roedder and Bodnar (1980) and Burrusss (1987). Perhaps the first important question that has to be considered is whether the inclusions are primary or secondary. The distinction is especially important in studies that aim at determining temperature history of different diagenetic stages. Hollister (1981) and Roedder (1984) have reviewed the various criteria for distinguishing between primary and secondary inclusions. In general, relatively large and isolated inclusions and clusters of such inclusions, or inclusions that outline growth zones of crystals are considered primary whereas a line or trail of relatively small inclusions crossing crystal boundaries are considered secondary (figs. 10, 11).

One of the more serious problems involves natural "re-equilibration" (Bodnar and Bethke, 1984; Goldstein, 1988) of inclusions since original entrapment. This may happen when the host mineral is subjected to elevated temperature as a result of subsidence. The pressure inside the inclusion may increase and equal or exceed the external pore-fluid pressure on the mineral and depending on the mechanical strength of the host mineral, the overpressure inside the inclusion may cause either brittle fracture of the inclusion wall (decrepitation) resulting in leakage of the inclusion-fluid, or permanent, plastic deformation (stretching) of the wall without leakage of the fluid. In either case the "re-equilibrated" inclusions give T_h that do not reflect their original trapping temperatures.

The main problem in identifying stretching is that there is seldom any physical sign that the observer can use to determine if an inclusion has been stretched. Even in experimental studies (Bodnar and Bethke,1984; Prezbindowski and Larese, 1987), volume changes of inclusions due to overheating are so small that these cannot be measured directly, and that stretching has occurred is inferred only from measured increase in T_h of the inclusions after reheating. Decrepitation is generally more easily identifiable, although some "healed hydrofractures" may defy microscopic identification (Goldstein, 1986). Experimental decrepitation of inclusions in natural quartz crystals under various effective stress conditions has recently been described by Vityk et al. (1994).

Another problem in fluid-inclusion study is estimating the pressure at the time of inclusion formation. Homogenization temperature usually gives a minimum estimate of the trapping temperature (T_t) of an inclusion, and to derive the true T_t a "pressure correction" is necessary for which one has to know the original pressure. Roedder and Bodnar (1980), Shepherd et al. (1985), and Burrusss (1987) have discussed the various methods and difficulties of pressure correction. In most cases, an independent estimate of paleopressure at the time of trapping of the inclusion is required. If the pressure is known, then the T_t can be obtained by using a combination of published P-V-T-X phase diagrams (Haas, 1976; Potter and Brown, 1975, 1976). Potter (1977) provides simplified P-T diagrams for aqueous solutions of various salinities from which the "pressure correction" can be read directly. The salinities in Potter's (1977) diagrams can be determined from the final melting temperatures (T_m) of fluid inclusions in the same heating-freezing stage.

An independent estimate of paleopressure is, however, seldom available, especially when missing overburden is involved. Even when thickness of the missing overburden can be estimated, the problem of choosing between hydrostatic and lithostatic pressure remains (Roedder and Bodnar, 1980). A convenient method of determining paleopressure (therefore, trapping temperature) involves using two sets of coeval inclusions in the same mineral crystal containing fluids of different compositions, which were present as immiscible fluid at the time of trapping (Roedder and Bodnar, 1980). The two immiscible fluids commonly found in sedimentary rocks are oil and water. If inclusions separately containing oil and water with their respective vapor phases can be proven to have been coeval, then their isochores will intersect at the trapping pressure and temperature (Burruss, 1987; Videtich et al. 1988).

4.3.1 Fluid-inclusion methods used in this study

Figure 12 shows the sample types in which the fluid inclusions of this study were generally found. These include inter- and intra-particle cements as well as vug- and fracture-filling cements of various types.

Fluid-inclusion samples were prepared first by cutting chips from hand samples with a low speed hand saw. Doubly polished thin sections, about 200 microns thick, were made from the chips. From the adjacent part of a sectioned chip standard thin sections were made for petrographic examination. Three to seven samples from each location were prepared in this way. The fluid-inclusion thin sections were broken into several small pieces and studied in a USGS Heating-Freezing Stage (University of Arizona) equipped with a Nikon binocular microscope. Details of this instrument are given in Hollister et al. (1981).

Rate of heating during homogenization runs were initially maintained at about 2°C/minute, but later reduced to 0.5°C/minute when the homogenization temperatures were approached. The same rate was maintained during freezing runs. Homogenization temperature of each inclusion was measured 2 to 3 times, but the results did not vary by more than 0.3° C. Freezing measurements were taken only in relatively large and clear inclusions, and immediately after homogenization measurements. Two freezing runs per inclusion were carried out at times. The results occasionally varied, but not by more than 0.5°C. Salinities, determined from freezing run, although have no bearing on paleotemperature assessment, will have important implications on the origin of the inclusions and indirectly on stable isotope interpretations.

Whether the inclusions are primary or secondary was determined using criteria discussed by Roedder (1984). The vast majority of inclusions measured in this study appear to be of primary origin. Secondary inclusions are present, but owing to their general small sizes only a few of them could be measured. However, the T_h values of both types of inclusions from a sample location were treated together in the histograms and in calculations of modal values because the main goal of this study

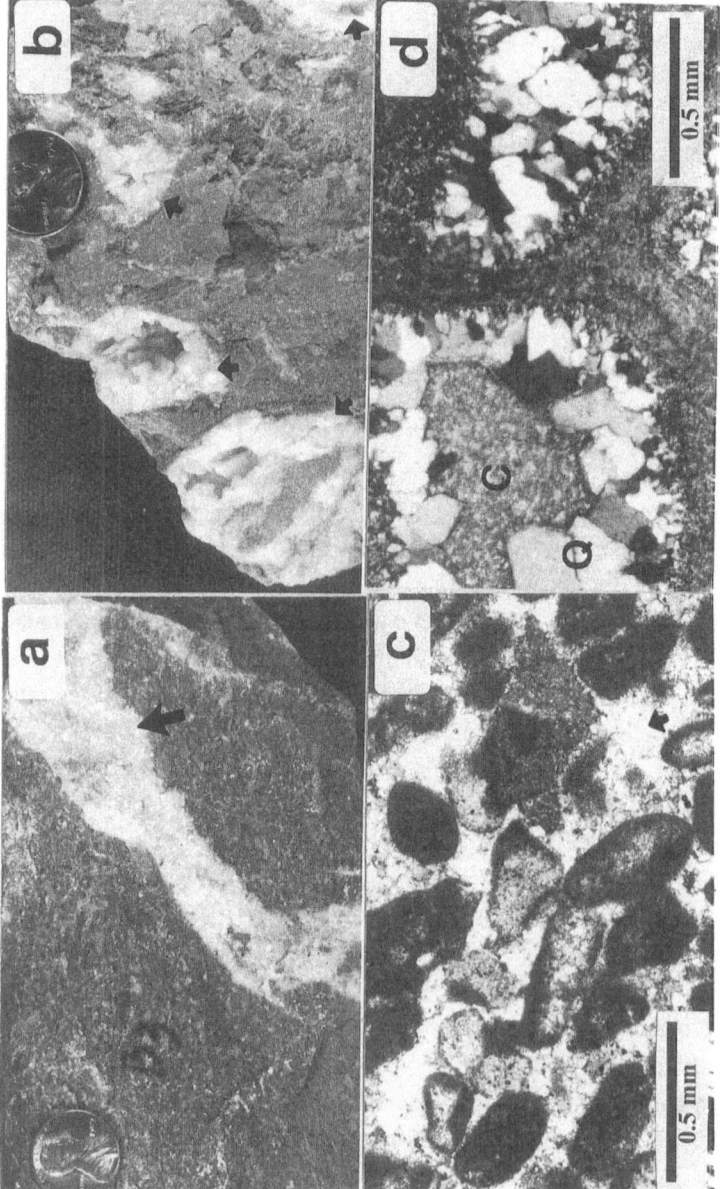

Figure 12: Photographs of hand samples and thin sections showing typical sources of fluid inclusion data: (a) calcite vein (marked by arrow) in crinoidal grainstone of Helderberg limestone, location 26; (b) vug-filling saddle dolomite (marked by arrows) in Middle Silurian Lockport dolostone, location 45; (c) interparticle calcite cement (marked by arrow) in peloidal grainstone of Upper Ordovician Black River limestone, location 87; (d) intra-particle quartz (Q) and sparry calcite (C) cement in coral cavities of Middle Devonian Onondaga limestone, location 7.

was to determine the maximum paleotemperatures experienced by the samples, not temperatures of various diagenetic stages.

On the basis of observations made in this study, it is difficult to determine how many of the inclusions have reequilibrated, although there are indications that some probably have (see section 5.1). Some precaution was taken to eliminate possible re-equilibrated inclusions from the study at the outset: inclusions with unusually large bubbles or "necked' and irregular, digitate-type shapes were not measured; clusters of very small inclusions arranged in a fashion suggesting decrepitation of a larger original inclusion were not measured even if some of them had gas bubbles in them.

For this study, however, re-equilibration may not be a serious problem, unless of course the reset values of T_h of most of the inclusions far exceeded their true trapping temperatures. Generally, however, the reset T_h only approaches the highest post-entrapment temperature, not exceed it. This is experimentally shown by Prezbindowski and Larese (1987). Barker and Goldstein (1990), based on their study of fluid inclusions in subsurface calcite cements in 46 diverse geologic systems that are presently at or near their maximum burial temperatures, also found excellent correlation between present formational temperatures and Th_{mode} (i.e., mean of T_h in the highest modal class) as well as Th_{mean}. Therefore, reset T_h may actually serve as 'maximum recording geothermometers' (Burrusss, 1987; Barker and Goldstein, 1990, Barker, 1991).

Pressure correction of the measured T_h values could not be performed because no reliable estimate of the original pressure (geobarometer) was available to the authors. The authors feel that the chances of error in assuming pressure values at the time of inclusion formation, especially when former depth of burial is not known, are too large to attempt any meaningful pressure correction. The technique of using coeval inclusions of two immiscible fluids (oil and water, for example) for pressure correction (see above) could not be used because hydrocarbon-bearing inclusions appeared to be very few in number and could not be positively identified. The paper by Roedder and Bodnar (1980) discusses the difficulties and pitfalls of oversimplified assumptions of paleo-geobarometric pressures. In this study, T_h therefore represents the lower limit of the true trapping temperature of an inclusion.

4.4 Clay diagenesis

Burial diagenesis of clay minerals provides a useful tool for determining thermal maturity of sediments. A number of diagenetic changes in clay minerals are observed with increasing burial. The most important mineralogical change involves the transformation of smectite to illite through a series of mixed-layer illite-smectite

(I/S) intermediaries (Perry and Hower, 1970; Weaver and Beck, 1971; Eslinger and Pevear, 1988; Moore and Reynolds, 1989).

The disappearance of smectite and corresponding appearance of illite with depth in sedimentary basins seems almost universal and is one of the fundamental reactions in clastic diagenesis (Abercombie et al., 1994). Young shales typically consist of 20-50% smectite, but in older and/or deeply buried rocks smectite may be virtually non-existent and illite may constitute more than 95% of the mixed-layer I/S (Eslinger and Pevear, 1988). A similar general decrease in the abundance of mixed-layer clays and kaolinite is observed in these rocks. At the same time, the percentage of chlorite increases.

Studies of several basins around the world show that I/S are typically randomly stratified to a depth of approximately 3050m (10,000 ft.) where, in some wells, an abrupt increase from about 40% to 70-80% illite is observed (Hower, 1981; Eslinger and Pevear, 1988). At these depths, the shale also assumes an ordered structure of alternating illite and smectite layers with extra illite randomly distributed. In other wells the transition from random to ordered inter- stratification is more gradual and in some wells, especially in rapidly buried sequences, the transition occurs at greater depths. In wells where information is available to depths greater than 6100m (20,000 ft.), the percentage of illite is generally found to be 90 or more (Eslinger and Pevear, 1988)

The nature of chemical reaction involved in smectite- illite transition in mixed layer I/S is not clear, or perhaps not unique, and several models have been proposed (see Moore and Reynolds, 1989, p. 144-157). The problem of understanding the transformation mechanism perhaps emanates from a lack of consensus on the nature of illite and smectite (for example, whether these should be treated as discrete minerals, particles of mixed composition, or solid solutions).

The chemistry of the transformation, however, requires that K and Al are incorporated into the newly formed illite. According to Hower et al. (1976), the original smectite layers remain intact and K and Al, derived from dissolution of K-feldspars, are substituted into the smectite structure to create illite. Silica, Mg and Fe released from smectite, if retained in the system, form chlorite and quartz. According to Boles and Franks (1979), Al is derived from dissolution and removal of part of the silica from the original smectite so that part of smectite is "cannibalized" and the total amount of clay is reduced. Nadeau et al. (1985) and Inoue et al. (1987) also proposed a stepwise dissolution and recrystallization reaction. The transformation process also causes expulsion of interlayer water from smectite, probably due to contraction of the layers resulting from an increase in charge as Al substitutes for Si (Eslinger and Pevear, 1988). According to Abercrombie et al. (1994), smectite-illite reaction occurs as a result of reduction in silica activity at the onset of quartz-precipitation.

The primary controls on smectite-illite transformation reaction are believed to be temperature, time, and rock and fluid chemistry. Of these, temperature has the most well-documented effect. High burial temperature apparently releases K necessary for illitization from coarse K-feldspar and mica (Perry and Hower, 1970; Weaver and Wampler, 1970; Weaver and Beck, 1971; Hower, 1981; Smart and Clayton, 1985). Pressure is probably not that important because it has been shown that pressures found in deep-burial environments are not high enough to dehydrate the interlayer water from smectite (Colten-Bradley, 1985; Eslinger and Pevear, 1988, p. 5 - 16).

The effect of time on smectite-to-illite reaction is controversial. In some experimental and field studies, reaction kinetics has been shown to be important (Eberl, 1978; Pevear et al. 1980; Ramsayer and Boles, 1986, Velde and Vasseur, 1992). However, Weaver's (1978) compilation of data, extending over an age range of 325 million years, shows temperature, not time, as the major control on illitization. It appears that kinetic factors are important in the lower temperature portion of the diagenetic sequence, particularly in younger (Tertiary) rocks (Eslinger and Pevear, 1988, p. 5-21). In older rocks, temperature is probably the most important factor.

Fluid and rock compositions are the other factors that play an important role in the smectite-to-illite reaction (Robertson and Lahan, 1981). However, little information is available on the composition of pore fluids in shale. It is at least certain that the shale must have sufficient K to satisfy the requirements of the smectite-illite reaction.

Smectite-illite transition has been used for measuring thermal maturity of sediments in at least two ways. The first one uses the percentage of illite in I/S clays to infer a range of temperature in which these percentages are observed in deep wells (Hoffman and Hower, 1979; Johnsson, 1986; Pollastro and Barker, 1986). For example, less than 50% illite in randomly stratified I/S indicates temperatures below 100° C; 60 - 80% illite layers indicate temperatures of 100 - 175° C, roughly coinciding with the "oil window"; and 85 - 95%, ordered illite layers indicate temperatures near or above 200° C (Hoffman and Hower, 1979; Eslinger and Pevear, 1988). Weaver et al. (1984) found 5% illite in I/S at about 210° C. McDowel and Elders (1980, cited in Weaver et al. 1984), in their work on Salton Sea geothermal system, show that the amount of illite in I/S of shale in question changes from 85 - 90% at 185° C to 95% at 210° C to pure illite at 275° C. Generally, however, the reaction rarely goes to 100% illite probably because K-feldspar, the main source of K, is exhausted before this stage is reached.

Besides percentage of illite in mixed layer I/S, illite-crystallinity index (I.C.) of clay samples has been used as a geothermometer or as a relative scale of thermal maturity of shales (Kubler, 1968; Frey, 1980; Kisch, 1980; Duba and William-Jones, 1983; Weaver and Boekstra, 1984; Guthrie et al. 1986). It is expressed as the width of the basal, 001, diffraction peak of illite at half its height. As ordering increases in I/S, variability of d-spacing decreases because the illite layers become more alike. In other words, the structure becomes more homogeneous and crystalline. This is reflected in the narrowing or sharpening of the 001 peak at 10Å. The peak can therefore be compared to a histogram, the smaller width of which represents greater homogeneity or crystallinity of the mineral (Duba and William-Jones, 1983).

A firmly established relationship between I.C. and temperature is yet to be known. However, I.C. has been used to mark boundaries between metamorphic zones in pelitic rocks. For example, I.C. of 0.42° 2θ is considered the boundary between "diagenetic zone" and "anchimetamorphic zone" or "anchizone", and I.C. of 0.25° 2θ the boundary between "anchizone" and "epizone" (Kubler, 1968; Weaver and Boekstra, 1984). Various attempts have been made to assign temperature ranges to these zones. On the basis of fluid-inclusion data from vein quartz in shale, Frey et al. (1980) found a minimum temperature of 200° C for the beginning of anchizone (i.e, I.C. equal to or less than 0.42° 2θ) and 270°C for the advanced stages of anchizone. Weaver et al. (1984) placed the beginning of anchizone at 280°C and the beginning of epizone at 360°C based on isotopic and bottom-hole temperatures of the Cambrian Conasauga shale of the southern Applachians (fig. 13).

4.4.1 Clay analysis methods used in this study

Shale samples were prepared for x- ray diffraction using methods described by Jackson (1979) and Moore and Reynolds (1989). The samples were washed, dried and, in case of outcrop samples, the surface layers were discarded by scraping. The samples were then crushed to mm-sized particles using an iron mortar and a pestle. The crushed samples were passed through a 0.2mm sieve to collect the -0.2mm powder. The powder was then treated successively with .1N HCl and 6% H_2O_2 solutions to remove carbonates and organic matters respectively. Subsequently they were dispersed in distilled water in graduated beakers and the - 20μm, -6μm, and -2μm size fractions were separated by gravity sedimentation and decantation using settling times for different grain sizes calculated by Jackson (1979). Water containing the -6μm and -2μm fractions were centrifuged to concentrate the clay. The concentrated water-clay mixture was sedimented on glass slides by an

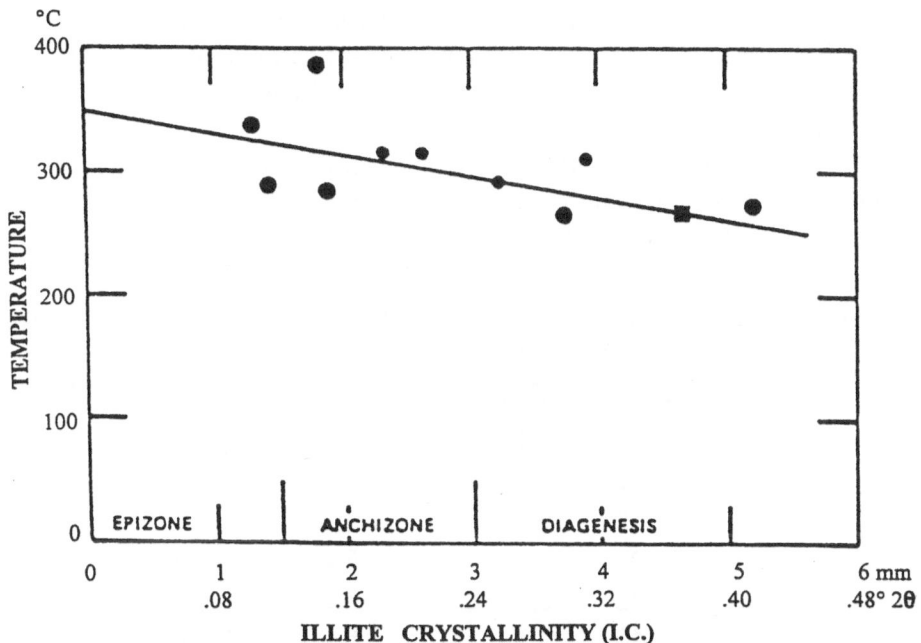

Figure 13: Temperature - illite crystallinity relationship diagram of
Weaver et al. (1984) based on their study of the Cambrian Canasauga
shale of the southern Appalachians.

eye-dropper. Two slides were made with the -2μm fraction for every shale sample.
Only a few slides with -6μm fraction were made. The slides were placed on
ceramic tiles in an oven and exposed to low temperatures (~40°C) where they dried
in about two hours.

The above procedure provided roughly oriented clay samples required to measure
the basal reflections by x-ray. The -2μm samples were used for x-ray analysis to
avoid measuring the detrital illites which are generally of larger size (Eslinger and
Pevear, 1988). Both slides with -2μm fraction were x-rayed to see if intra-sample
variation existed. One of the slides was then solvated with ethylene glycol by
dripping from an eye-dropper and drying for a brief period. The solvated sample
was then x-rayed again. Glycol-solvation is necessary for identification of certain
clay minerals, such as the smectite, which expands upon glycol treatment to its
specific basal spacing (Eslinger and Pevear, 1988, A-22).

The clay samples were analyzed by a RIGAKU X-ray diffractometer, DMAX IIB
with CuKa radiation of 2KW maximum intensity. For each sample, scanning was
done between 4° and 30° at a goniometer speed of 1°/minute and voltage of 40KV.
A 'peak finding program' automatically calculated the peak positions, peak
intensities, and the widths of the peaks at half of their heights. Width of the 10Å

peak found this way provided the I.C. (in °2θ) for each sample. Approximate percentage ranges of illite were calculated by using the positions of the 001/002 and 002/003 I/S peaks (table 2) after Moore and Reynolds (1989).

Table 2: Percentages of illite in mixed-layer illite/smectite based on positions of 001/002 and 002/003 reflections in X-ray diffractograms of glycolated clay samples. (From Moore and Reynolds, 1989).

% illite	001/002		002/003		°Δ2θ
	d (Å)	°2θ	d (Å)	°2θ	
10	8.58	10.31	5.61	15.80	5.49
20	8.67	10.20	5.58	15.88	5.68
30	8.77	10.09	5.53	16.03	5.94
40	8.89	9.95	5.50	16.11	6.16
50	9.05	9.77	5.44	16.29	6.52
60	9.22	9.59	5.34	16.60	7.01
70	9.40	9.41	5.28	16.79	7.38
80	9.64	9.17	5.20	17.05	7.88
90	9.82	9.01	5.10	17.39	8.38

In this study, percentage of illite in I/S was used to set limits of paleo-temperatures taken from the references cited above. For samples containing more than 90% illite - in fact, all samples fell under this category - it was inferred that the sample had experienced temperature of at least 200° C. To correlate I.C. with paleotemperatures Weaver et al.'s (1984) I.C. - temperature chart (fig. 13) was used. Although Weaver et al.'s (1984) study is from a different suite of rocks, use of their chart at least provided some additional constraints on paleotemperatures of the studied rocks.

4.5 Organic maturation

Organic materials undergo various physical and chemical changes after burial. These changes are influenced by temperature, pressure and geologic time, but the relative importance of these factors is a subject of controversy. 'Maturity' is a measure of post-depositional alteration of organic matter. There are numerous methods of assessing maturity of organic matter in sediments (Heroux et al. 1979;

Crelling and Dutcher, 1980), but 'vitrinite reflectance' (Ro) is most widely used. It is the primary standard by which coal is ranked (Stach et al. 1982) and is routinely used to evaluate hydrocarbon source rocks.

Vitrinite is the principal maceral type in coal and makes up 50 - 90% of most North American coals (Crelling and Dutcher, 1980). It is generally the most abundant constituent of sedimentary kerogen (Dow and O'Connor, 1982). Most vitrinite macerals are derived from the cell-wall material or woody tissue of plants. Vitrinite is formed in the diagenetic environment by humification of this material.

With increasing maturity, the aromatic lamellae of vitrinite become more ordered resulting in a systematic increase in optical reflectivity (Dow and Connors, 1982). This reflectivity (vitrinite reflectance) is measured as the percentage of light (ranges between 0.3 and 5%) vertically incident on a polished vitrinite surface, that is reflected back to the viewer.

Although vitrinite reflectance (Ro) measuring procedures were first developed for coal petrology, these have been widely used in the study of disseminated organic matter, or kerogen, in rocks that do not contain coal (Castano and Sparks, 1975; Cardott and Lombard, 1985; Houseknecht and Mattews, 1985; Guthrie et al. 1986; Clayton, 1989; Connolly, 1989, Johnsson, 1993). Separation and measuring techniques of vitrinites of kerogen are discussed by Dow and Connors (1982) and Stach et al. (1982).

As many Ro measurements as possible are taken from a rock sample - usually at least 50 are recommended (Castano and Sparks, 1975). A high number of Ro readings is necessary in order to separate out 'primary' and 'recycled' vitrinites. Although it is not always possible to separate them, vitrinites recycled from erosion of older rock units generally tend to show higher Ro. Also, vitrinites in highly weathered or oxidized samples show higher Ro. In core samples, another source of non-indigenous vitrinite is from caving. Vitrinites from this source usually show lower reflectance values. Only the primary or indigenous vitrinites can be used for maturity analysis. Despite all precaution, it is often difficult to identify the primary vitrinite population from Ro histograms when several populations are present. To overcome this problem, it has become a common practice to use mean Ro to represent organic maturity.

How vitrinite reflectance data can be best used to estimate paleotemperatures experienced by a given rock has remained unresolved. It is generally agreed that temperature is the dominant factor influencing vitrinite reflectance, and that the effect of pressure is negligible in most cases (Crelling and Dutcher, 1980; Barker, 1983). The source of contention is in the role of geologic time which has split the available methods of assessing maturity of sediments into two groups: (A) those that propose geologic time as an important factor (Lopatin, 1971; Hood et al. 1975;

Waples, 1980), and (B) those that propose geologic time as having only a limited importance (Barker, 1983, 1988; Price, 1983; Barker and Pawlewicz, 1986).

Among the time-dependent methods, perhaps the most widely used method involves the correlation of vitrinite reflectance with a calculated time-temperature index (TTI) based on the assumption that chemical reaction in organic materials follows a first-order kinetics (i.e., obeys Arrhenius Equation), and the reaction rate doubles with every 10° C increase in reaction temperature (Lopatin, 1971; Waples, 1980; Horvath et al., 1988; Sweeny and Burnham, 1990). Morrow and Issler (1993) have reviewed the various time-dependent methods.

Barker (1983) and Price (1983), however, questioned the validity of time-dependent models of organic maturation on both theoretical and empirical grounds. According to them, there is no evidence from the natural system that petroleum generation - maturation reactions have first-order reaction kinetics; on the contrary, laboratory experiments show that these reactions follow multiple-order kinetics. Price (1983) presented mean vitrinite reflectance data from deep wells of several sedimentary basins of the world which show a strong correlation with ambient temperatures but not with burial times for any temperature interval. Full organic maturation at a given temperature is believed to be achieved within 1 million years (and as short as 1000 - 10,000 years in hot geothermal systems) after which reaction time has no effect. That the geologically older sediments often show higher organic maturity than younger sediments at the same burial depths can be explained, according to Price (1983), by the greater chance the older rocks have being affected by major thermal events. Based on their findings, Barker (1983), Price (1983), and Barker and Pawlewicz (1986) propose that vitrinite reflectance can be used directly as an absolute geothermometer. The following linear regression equation of Barker and Pawlewicz (1986), based on data from 35 geologic systems, shows the relationship between maximum burial temperature (T_{mb}) and mean vitrinite reflectance (R_m):

$$\ln (R_m) = .0078T_{mb} - 1.2$$

In a later publication, Barker (1988), however, shows that although the time-dependent and time-independent models use different approaches, they show comparable fit with data from natural systems. This is because, he explains, the time-dependent models such as that of Lopatin (1971) are heavily weighted toward reaction occurring near the maximum burial temperature, because these models rely on the reaction rate doubling with every 10°C temperature increase.

Since land plants first appeared only in the latest Silurian time, vitrinites are generally absent in pre- Devonian rocks. Bulk of the organic matters in these rocks are exinites (spores, pollens, cutinites) and amorphous sapropelic substances (Staplin, 1975). It was observed by Teichmuller (1952) that in rocks containing both coal and exinite it was possible to roughly estimate the rank of coal on the basis of color and reflection changes in exinite. That is, exinite changes color from yellow through brown to black in response to increasing temperature. Later, a numerical scale called Thermal Alteration Index (TAI) was introduced (Staplin, 1969, 1975) to represent color alterations of exinite and for the purpose of correlation with other maturation indicators. Direct relationship between TAI and temperature has not been established probably because TAI is based largely on a qualitative property, color. However, temperatures can be estimated from TAI via its equivalent Ro (table 3) following methods described above.

In general, TAI values do not approach the level of precision provided by Ro (Bustin, 1989; Morrow and Issler, 1993). Tissot et al. (1980) discussed several problems of TAI measurements and usefulness. These are (1) the subjective nature of color determination and calibration problems between laboratories, (2) dependence of color on the original nature of plants involved and thickness of particles inspected, (3) color alteration by early diagenetic processes, such as oxidation, and (4) hydrogen- poor particles with a high aromacity are darker than hydrogen-rich particles with a low aromacity, even if they have the same thermal history.

Table 3: Interconversion of thermal alteration index (TAI) and vitrinite reflectance values (Ro). (From Geochem Laboratories, Houston, 1988).

Thermal alteration index (TAI)	Descriptive maturity terminology	Vitrinite reflectance (% Ro)
1.00	Immature	
1.10	--	
1.20	--	
1.30	--	0.30
1.40	--	0.33
1.50	--	0.37
1.60	--	0.40
1.70	--	0.43
1.80	Moderately immature	0.45
1.90	--	0.48
2.00	--	0.50
2.10	--	0.55
2.20	Moderately mature	0.60
2.30	--	0.70

(**Table 3** continued)

2.40	--	0.75
2.50	--	0.80
2.60	Mature	0.90
2.70	--	0.93
2.80	--	0.95
2.90	--	0.98
3.00	--	1.00
3.10	--	1.13
3.20	--	1.25
3.30	--	1.38
3.40	--	1.40
3.50	--	1.63
3.60	Very mature	1.75
3.70	--	1.88
3.80	--	2.00
3.90	--	2.13
4.00	--	2.25
4.10	--	2.38
4.20	Severely altered	2.50
4.30	--	2.75
4.40	--	3.00
4.50	--	3.50
4.60	--	4.00
4.70	--	
4.80	--	
4.90	--	
5.00	Metamorphosed	4.50

4.5.1 Organic maturation methods used in this study

Samples were measured by Geochem Lab of Houston, Texas for Ro and TAI. In Devonian shale samples both Ro and TAI were determined, whereas in Devonian carbonates only TAI was measured. In pre-Devonian shales, because of reasons mentioned above, only TAI was measured. No TAI measurements were made on pre-Devonian carbonates when it was found that the TAI data from Devonian carbonates were quite inconsistent with other data. From mean Ro (and Ro-equivalents of TAI) data of samples of individual locations the maximum burial temperatures were calculated using Barker and Pawlewicz's (1986) equation.

4.6 STABLE ISOTOPES

Measurement of oxygen and carbon isotopic composition of various components of sedimentary rocks has become a common practice in modern diagenetic studies (see Arthur et al. 1983; Schneiderman and Harris, 1985; and Shukla and Baker, 1988 for references). There are two principal uses of stable isotope studies: (1) to determine the nature (marine, meteoric, hypersaline, hydrothermal etc.) of the diagenetic fluids, and (2) to determine the formational temperatures of the minerals in which they are measured.

The usefulness of isotopes for the above purposes rely on "isotopic fractionation" or partial separation of isotopes (owing to their different weights) which can occur during physical and chemical reactions. The fractionation factor for the $CaCO_3$-H_2O system in which the CO_3^{2-} ions in calcite are in equilibrium with the oxygen and carbon in water is given by

$$\alpha = \frac{(^{18}O/^{16}O)\ CO_3^{2-}}{(^{18}O/^{16}O)\ H_2O}$$

where $(^{18}O/^{16}O)CO_3^{2-}$ is the ratio of ^{18}O to ^{16}O in the carbonate ions and $(^{18}O/^{16}O)H_2O$ is their ratio in water. If the oxygen isotopes behave identically chemically, α would equal 1. The same type of relationship also applies for carbon $(^{13}C, ^{12}C)$ isotopes.

The value of "α" varies with temperature which makes possible the use of oxygen-isotope data for paleotemperature determination. Also, for a given temperature, the $^{18}O/^{16}O$ of carbonate should vary directly with the same of water. This makes it possible to infer the nature of the carbonate- precipitating fluid by analogy with isotopic compositions of modern meteoric, marine and hypersaline waters.

In oxygen and carbon isotopic analysis of a sample, it is customary to present the data in terms of difference in $^{18}O/^{16}O$ or $^{13}C/^{12}C$ (expressed as $\delta^{18}O$ or $\delta^{13}C$) between the sample and a common standard rather than the absolute isotopic ratios. The δO^{18} is mathematically expressed as

$$\delta^{18}O = \frac{(^{18}O/^{16}O_{sample} - {}^{18}O/^{16}O_{standard})}{(^{18}O/^{16}O_{standard})}\ 1000$$

The standards used are generally either PDB (belemnite of Pedee Formation) or SMOW (Standard Mean Ocean Water).

The temperature dependence of oxygen-isotope fractionation between water and some $CaCO_3$ phases are well known. The calcite-water fractionation determined from various experimental biogenic and inorganic reactions as well as theoretical calculations agree quite well at all temperatures except below 10°C (Anderson and Arthur, 1983, p.1-26). Friedman and O'Neils' (1977) equation for calcite-water, for example, is

$$10^3 \ln\alpha = 2.78 \times 10^6 T^{-2}\ (°K) - 2.89$$
where $10^3 \ln\alpha = \delta^{18}O_{calcite} - \delta^{18}O_{water}$ and
T = temperature of precipitation of calcite

The isotopic geochemistry of dolomite is not well understood and has been extensively written about (Zenger et al. 1980; Anderson and Arthur, 1983; Machel and Mountjoy, 1986; Shukla and Baker, 1988). The problem stems from the fact that well-ordered dolomite has not been synthesized at low temperatures and therefore the equilibrium oxygen isotopic fractionation between dolomite and water at sedimentary temperatures is not known (see Land, 1980). However, results of high-temperature experiments have been extrapolated to sedimentary temperatures (Northorp and Clayton, 1966, O'Neil and Epstein, 1966) and Fritz and Smith (1970) have used their low-temperature synthesis of poorly ordered "protodolomite" to do the same. Fritz and Smith's (1970) equation for dolomite-water can be written as

$$103\ln\alpha = 3.2 \times 10^6 T^{-2} \ (°K) - 3.3$$
where $10\ln\alpha = \delta^{18}O_{dolomite} - \delta^{18}O_{water}$ and
T = temperature of precipitation of dolomite

In the above two equations, $\delta^{18}O$ of calcite and dolomite are obtained by measuring the isotopic compositions of the samples, but $\delta^{18}O_{water}$ and temperature (T) are both unknown and either of the two has to be known to calculate the other. It becomes difficult to determine temperature by this method because there is no reliable method of determining the isotopic composition of the water from which calcite or dolomite precipitated: fluid inclusions in sedimentary rocks provide very little fluids for direct measurements. One can use a known range of $\delta^{18}O_{water}$ values from modern formation waters, but the measured range in carbonate rocks is much too large; for example, from -16 SMOW (Hitchon and Friedman, 1969) to +29 SMOW (Choquette and James, 1987) to be useful for paleotemperature measurements. There is a better chance of determining $\delta^{18}O_{water}$ using the above relationships because there are several methods, including the ones used in this study, to obtain a rough estimate of the temperature of precipitation.

Although, theoretically $^{13}C/^{12}C$ ratio of calcite samples could be used to determine temperature of precipitation, the fractionation of carbon isotopes between calcite and water has been found to be rather insensitive to temperature changes (Anderson and Arthur, 1983, p.1-80). This is probably because (1) the amount of fractionation between ^{13}C and ^{12}C is less than that between ^{18}O and ^{16}O, and (2) the mass of dissolved carbon in natural water is much less than that of oxygen and, as a result, carbon isotopes are more affected by local chemical processes. Relative contribution from dissolved organic matter may also change the local carbon content of water enormously (Anderson and Arthur, 1983).

Another major problem encountered in isotope studies is with sampling. Several generations of cements with different isotopic signatures may fill the same interstice in a sedimentary rock (Scholle and Halley, 1985; Kaufman et al. 1990). These cement zones are commonly a few tens of microns or less in thickness and the conventional methods of drilling cements out of a sample by using a dental drill or needle is too coarse. Thus, the cement specimens used in isotope analysis usually represent bulk samples and the data obtained are an average for different generations of cements.

4.6.1 Stable-isotope methods used in this study

Considering the limited usefulness of stable-isotope techniques for paleotemperature measurements, oxygen-isotopic analysis was restricted to 50 samples from Devonian and Silurian carbonates. The Ordovician and Cambrian carbonates, which were sampled in a later field season, were not analyzed for isotopes.

Samples were obtained mainly from vug- and fracture-filling cements by using a dental drill on polished slabs or chips: thus, most of our isotope samples probably included cements of more than one generation. The slabs or chips were cut opposite to fluid-inclusion thin sections. Sometimes, samples were obtained by breaking cements from a thin section with a needle under a microscope after fluid-inclusion work has been done on the thin section. Mass spectrometric work on the samples was conducted by Geochron Laboratories, Cambridge, Massachusetts. We used our oxygen isotope data to mainly calculate the possible $\delta^{18}O_{water}$ values for several of our samples. The mean T_h of the samples was used in the above equations as temperature of precipitation..

5 ANALYTICAL RESULTS

5.1 Fluid inclusions

Homogenization temperatures (T_h) of the studied samples show wide ranges in most of the study locations (fig. 14). A large range of T_h in rocks of a single study location may represent fluid trapping under different temperatures through their geologic history; however, re-equilibration from post-entrapment leaking and stretching may also result in an apparent large range (Goldstein, 1986, 1988). It is uncertain to what extent re-equilibration of the inclusions has taken place in the studied rocks. However, in some samples the presence of inclusions only a few microns to few tens of microns apart often showing T_h-difference of more than 100°C and of inclusions with unusually large, immobile vapor bubbles suggest that some re-equilibration has taken place in these samples (see Goldstein, 1988). In some samples, presence of inclusions with large, immobile vapor bubbles- which are believed to have leaked - in close proximity with normal-looking inclusions perhaps also indicate that "stretching" of these latter inclusions "without leakage" might have taken place. However, as discussed above, reequilibrated T_h may still be useful as recorders of maximum burial temperatures.

In this study, in order to assess the maximum T_h (Th_{max}) at a location, the average of all the T_h in the highest 20° C interval of the T_h histograms of individual locations has been used. Although somewhat arbitrary, this interval was chosen because at least

52

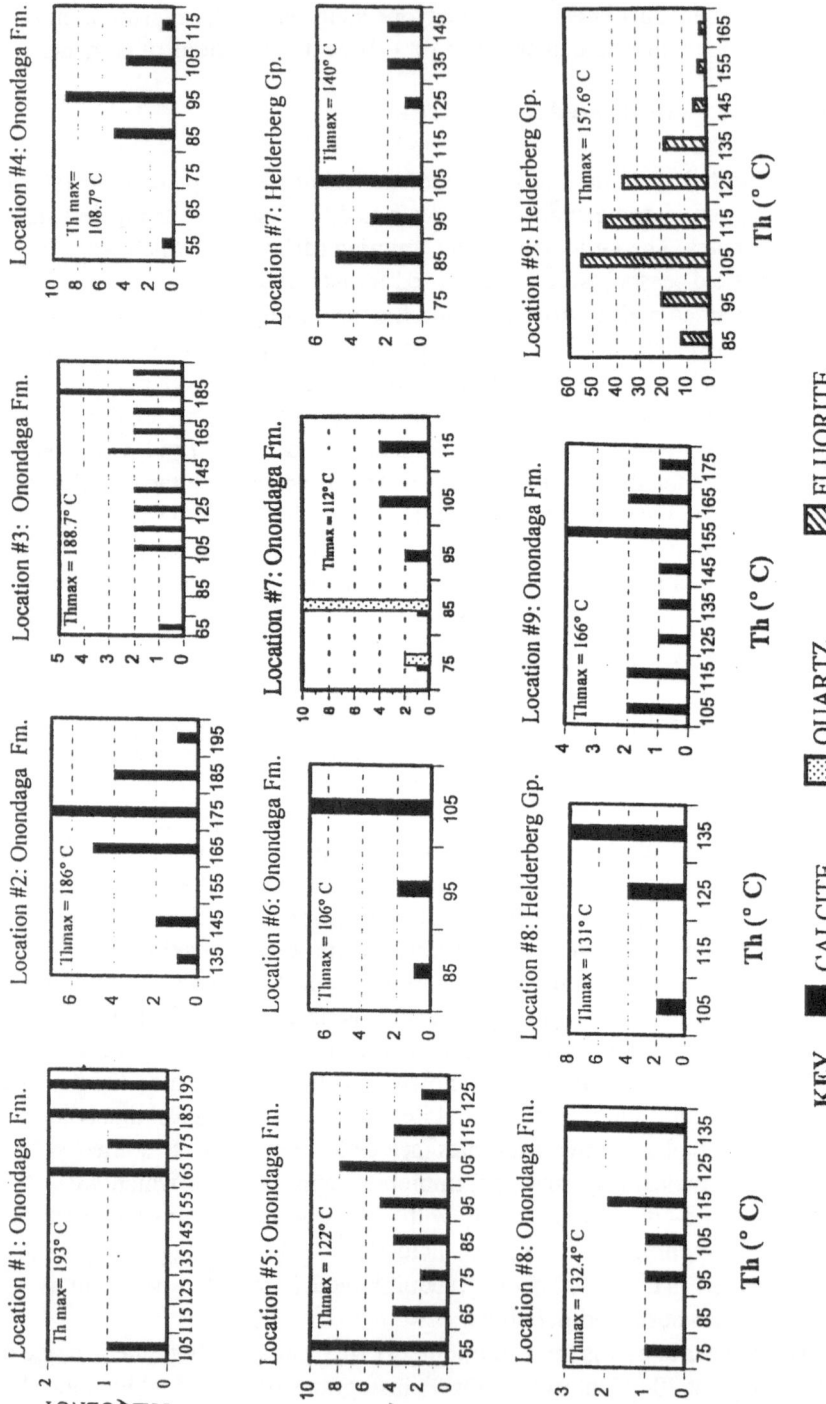

Figure 14: Histograms of fluid - homogenization temperatures (T_h) from individual locations. Th_{max} represents the mean of the T_h values in the top 20° C interval of each location.

53

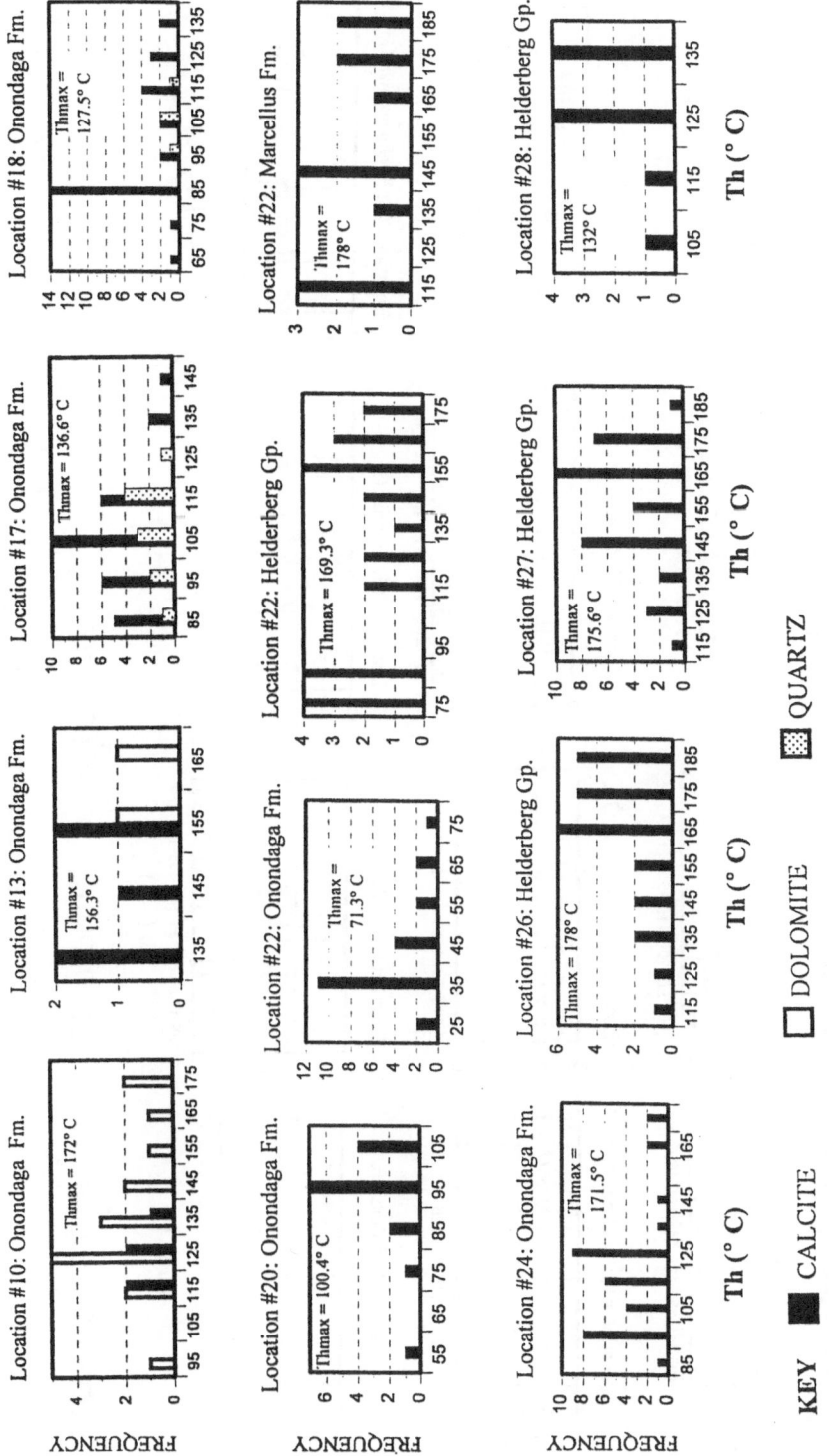

Figure 14 (continued)

54

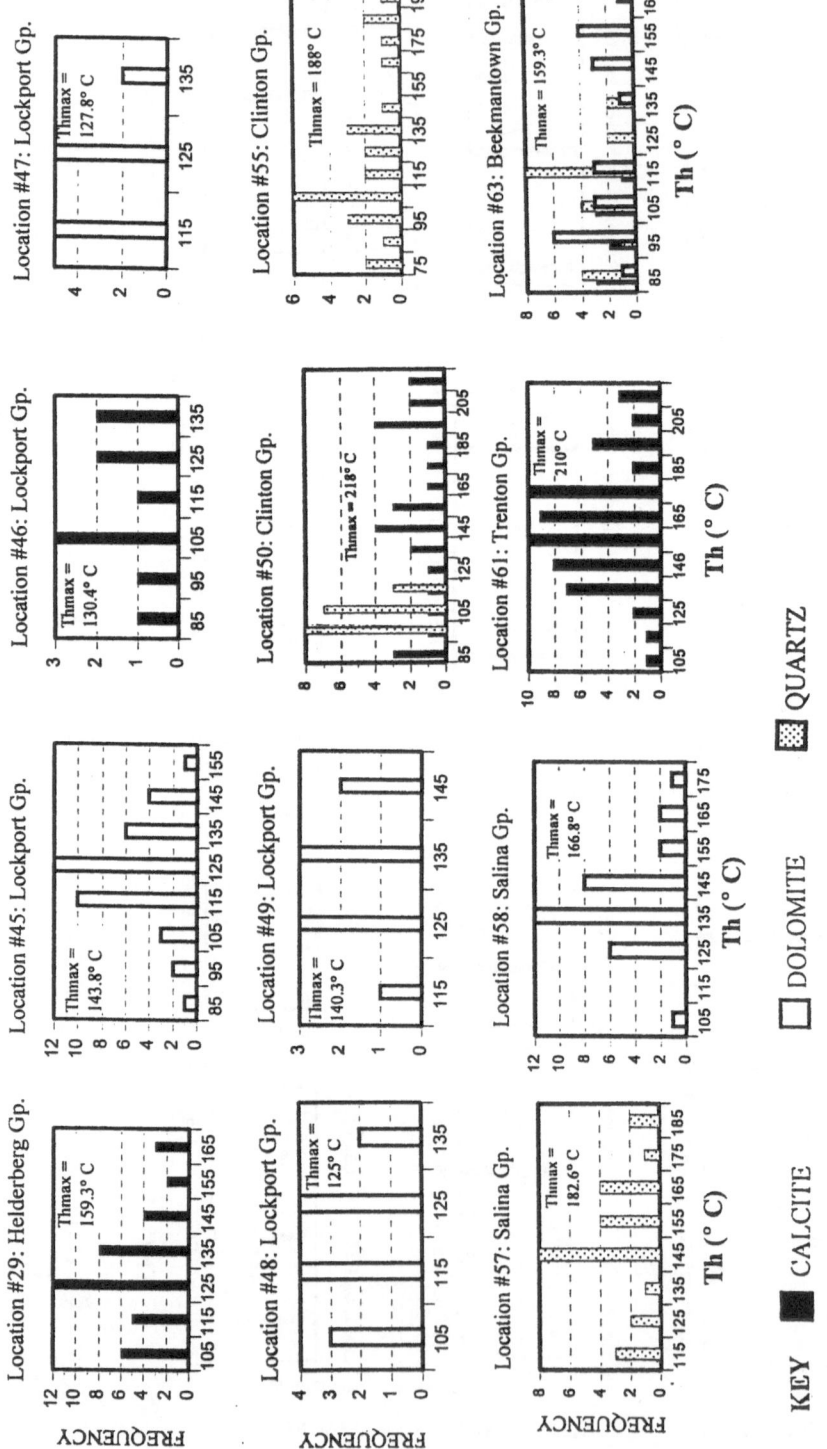

Figure 14 (continued)

55

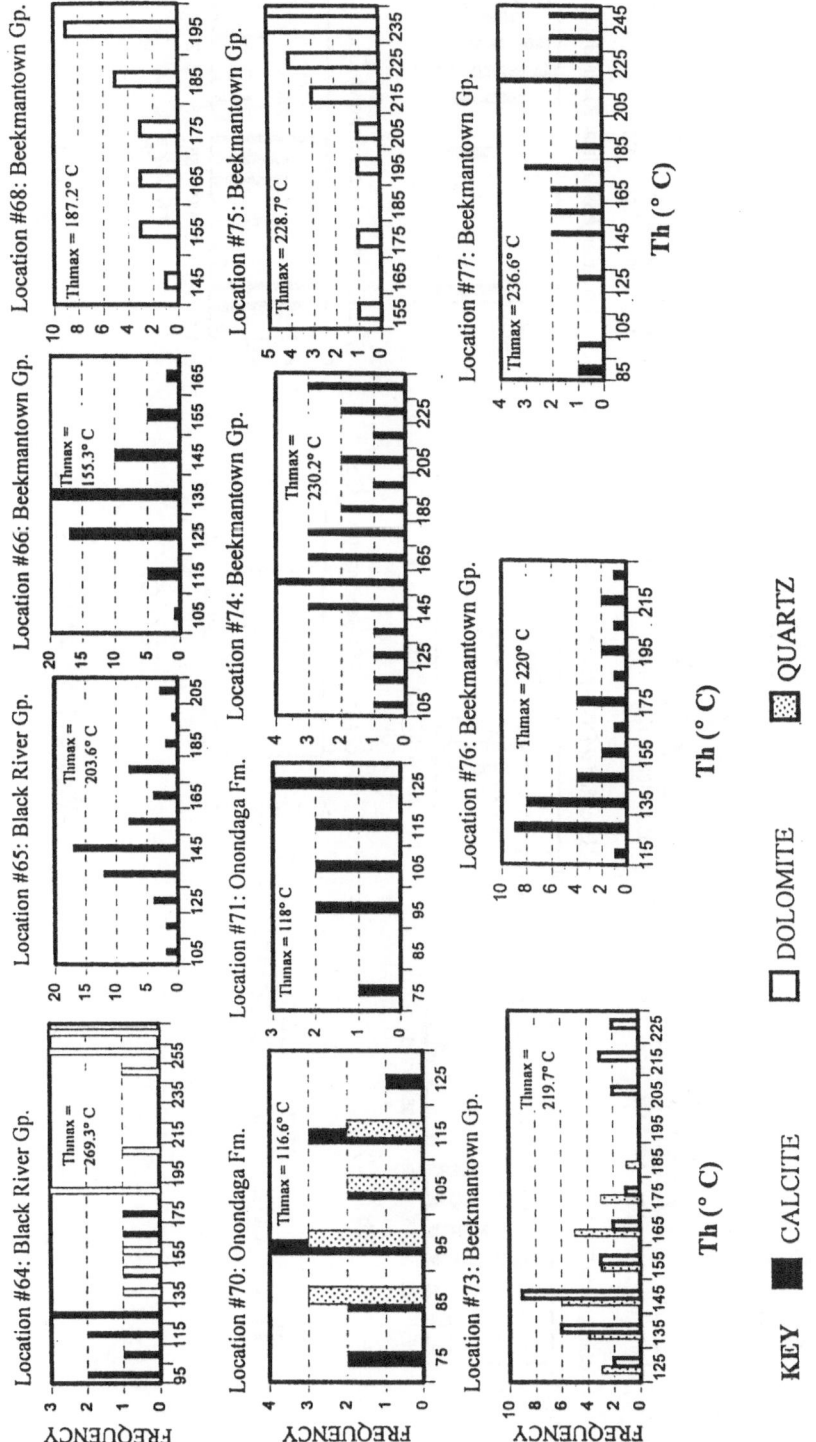

Figure 14 (continued)

56

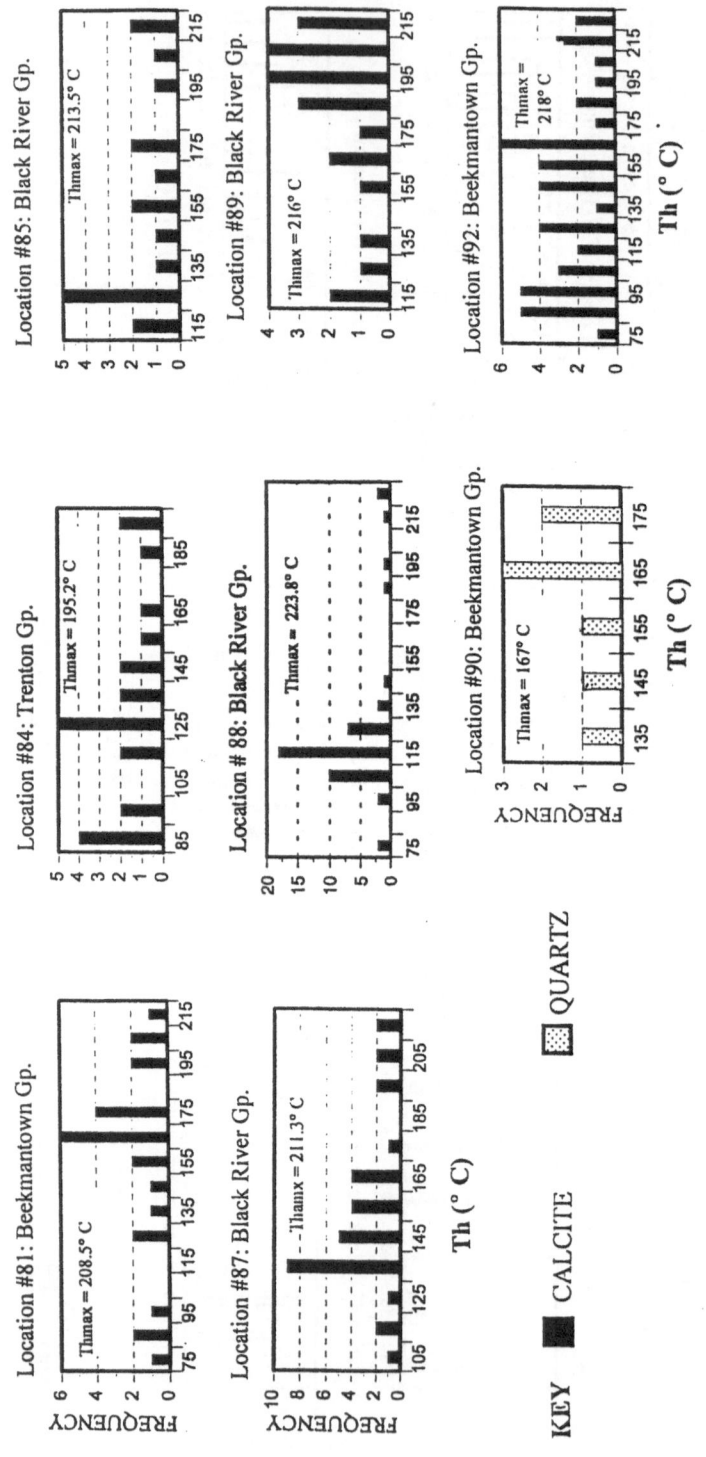

Figure 14 (continued)

three T_h readings were available in this interval at every study location and it was decided that this average value was more reliable and conservative than a single highest T_h. The Th_{max} values of individual study locations have been presented in figure 14 as well as table 7.

In figure 15, the measured T_m and the corresponding T_h values of calcite, dolomite, and quartz cements in the studied rock units have been plotted. Although the dataset is probably not large enough to allow definite conclusions to be made, it appears that the inclusion fluids in dolomite as well quartz cements generally had high T_m and therefore high salinity (14 - 30 wt% NaCl). T_m in calcite cement, on the other hand,

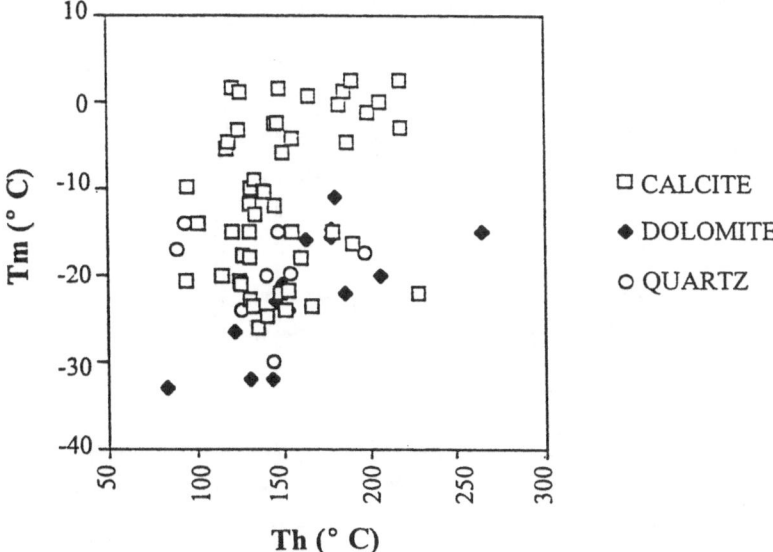

Figure 15: Fluid-homogenization (Th) vs. ice-melting temperature (Tm) diagram for inclusions in calcite, dolomite, and quartz cement.

ranges between +3 to -26° C indicating fluids ranging from 0 to about 27 wt% NaCl. The few positive T_m in figure 15 probably indicate meteoric waters, although it could be argued that these T_m are actually incongruent melting temperatures of hydrohalite crystals in fluids of high initial salinity (Roedder, 1962). However, T_h, which reflect the minimum trapping temperature of the inclusions show an equally wide range in all types of cements and these temperature do not systematically vary with T_m (e.g., salinity). The data also fails to show any recognizable spatial trend within the study area.

5.2 Clay diagenesis

Typical clay minerals in the studied samples are illite and chlorite possibly with minor amounts of kaolinite, and smectite in mixed layer illite-smectite (I/S) structure (fig. 16). The presence of kaolinite may be indicated by the peaks at 12.5° 2θ (7.05Å)

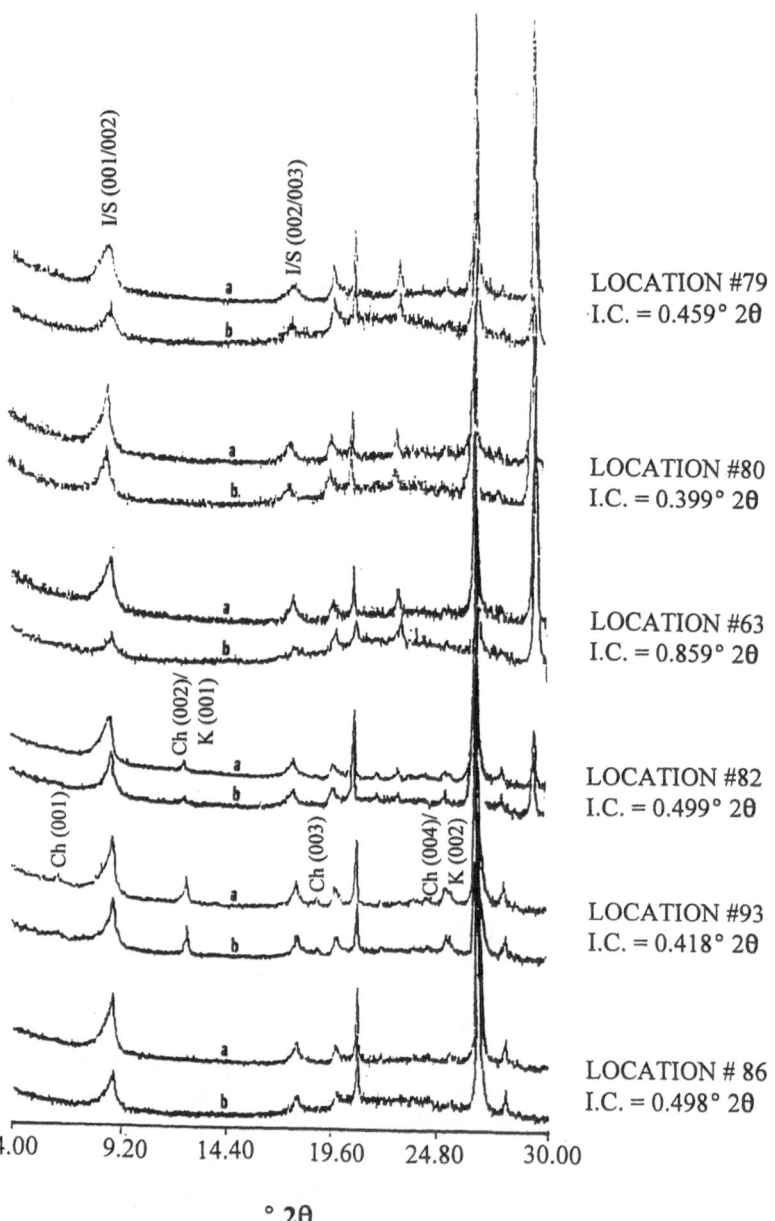

Figure 16: Representative X-ray diffractograms of untreated (a) and glycolated (b) clay samples showing important clay peaks. I/S = illite/ smectite, Ch = chlorite, K = Kaolinite.

and 25.2° 2θ (3.53Å) which are also the positions of chlorite 002 and 004 peaks respectively. Experimental procedures required to separate out kaolinite and chlorite peaks (Moore and Reynolds, 1989) at these positions were not undertaken in this study. However, lack of appreciable shift in the positions of 001/002 or 002/003 I/S peaks between glycolated and non-glycolated clays of the same samples (fig. 16) indicate that expandable smectite is virtually non-existent in these samples. Absence of the 001 peak of smectite at 5.2° 2θ in all samples also points against the presence of much free smectite. Difference in peak positions of 001/002 and 002/003 I/S peaks shows that clay samples from every study location contains more than 90% illite in mixed-layer I/S (table 4). As previously discussed, these characteristics are indicative of deep-burial diagenesis of clay and such high percentages of illite have commonly been measured in rocks subjected to temperatures of 200° C or more.

The illite-crystallinity index (I.C.) of the studied samples ranges from 0.317° to 0.859° 2θ (table 4). Most of the measured I.C. values straddle the boundary between the "diagenetic zone" and "anchizone" (0.42° 2θ). Approximate temperatures calculated from individual I.C. following Weaver et al. (1984) are presented in table 4. It is seen that the temperatures derived from I.C. are consistent with the temperature limits of 200° C indicated by percentages of illite in the same samples; only in location 63, I.C. indicates temperature lower than 200 °C.

5.3 Organic maturation

Results of vitrinite reflectance (Ro) measurements on shale samples from ten locations are shown in Ro-histograms of figure 17 as well as in table 5. TAI measurements from forty three locations, performed on both carbonate and shale samples - including the Ro samples - are also presented in table 5. The table shows the temperatures calculated from mean Ro and Ro-converted TAI values on the basis of Barker and Pawlewicz's (1986) equation.

The organic matter in the studied samples comprises amorphous sapropelic, herbaceous (spore-pollen), woody structured, and inertinitic components mixed in various proportions. Some of the Ro samples (locations 3, 22, 25, 36) do not contain any woody-structured components and the reflectance of these samples are probably a function of their inertinite content. The number of available R_o measurements at some locations (for example, 25 and 36) are probably too few to be of much value. The TAI values of the studied samples vary widely, from 2.2 to 3.9. These variations do not correlate with the percentages of the above organic matters in them. Also, the TAI values do not match well with the Mean Ro (%) of the same samples (when both are available from the same sample) and temperatures obtained from Mean Ro (%) are generally higher of the two (see section 6).

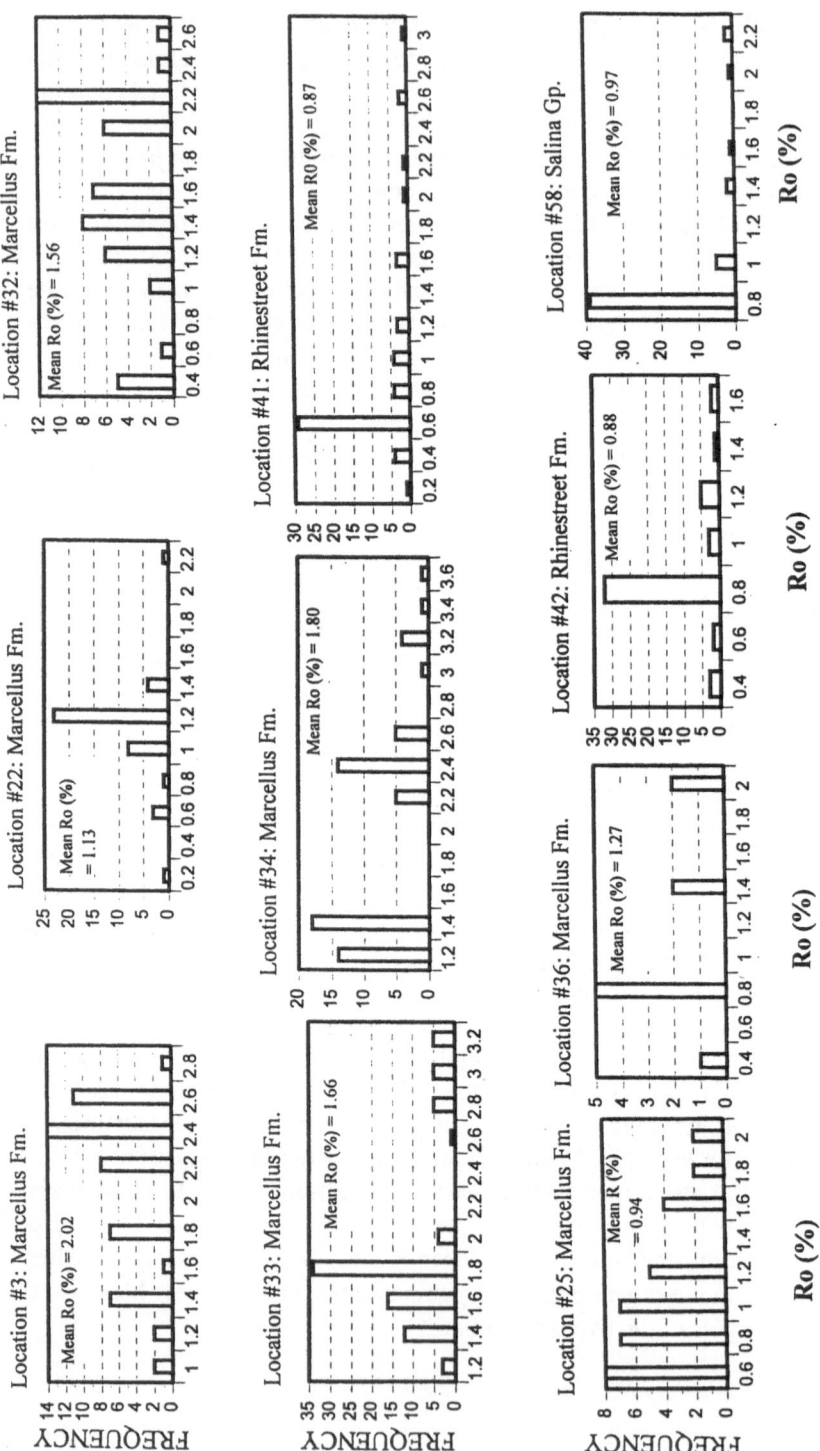

Figure 17: Vitrinite reflectance (Ro) histograms of shale samples from ten locations.

Table 4: Table summarizing the results of clay diagenesis study. Temperatures have been calculated following Weaver et al. (1984).

Location	Rock unit	Position of I/S 001/002 peak (°2θ)	Position of I/S 002/003 peak (°2θ)	Δ (°2θ)	% of Illite	I.C. (°2θ)	Approx. Temp. (° C)
3	Marcellus Fm.	8.74	17.76	9.02	90 - 100	0.320	280
11	Marcellus Fm.	8.78	17.78	9.00	90 - 100	0.439	250
22	Geneseo Fm.	8.86	17.78	8.92	90 - 100	0.439	250
	Marcellus Fm.	8.74	17.68	8.94	90 - 100		
		8.70	17.66	8.96	90 - 100	0.399	260
25	Marcellus Fm.	8.84	17.78	8.94	90 - 100	0.398	260
32	Marcellus Fm.	8.86	17.78	8.92	90 - 100	0.358	275
33	Marcellus Fm.	8.86	17.74	8.88	90 - 100	0.399	260
36	Marcellus Fm.	8.86	17.30	8.44	90 - 100	0.317	280
39	Geneseo Shale	8.80	17.68	8.88	90 - 100	0.439	250
41	Dunkirk Fm.	8.82	17.72	8.90	90 - 100	0.378	260
	Rhinestreet Fm.	8.80	17.74	8.94	90 - 100	0.357	275
42	Rhinestreet Fm.	8.86	17.74	8.84	90 - 100	0.599	225
51	Vernon Shale*	8.92	17.80	8.88	90 - 100	0.439	250
52	Vernon Shale*	8.90	17.70	8.80	90 - 100	0.458	245
59	Medina Shale	8.94	17.82	8.88	90 - 100	0.599	225
63	Utica Shale	8.84	17.76	8.90	90 - 100	0.859	170*
64	Utica Shale	8.88	17.76	8.88	90 - 100	0.359	275
79	Utica Shale	8.84	17.76	8.92	90 - 100	0.459	245
80	Utica Shale	8.84	17.76	8.92	90 - 100	0.399	260
82	Utica Shale	8.86	17.76	8.90	90 - 100	0.499	240
86	Utica Shale	8.84	17.76	8.92	90 - 100	0.498	240
93	Utica Shale	8.88	17.76	8.88	90 - 100	0.418	255

◆ = However, percentage of illite in this sample indicates temperatures of 200° C or more.

* = of the Salina Group

Table 5: Summary table for organic maturation analysis. Temperatures have been calculated from mean Ro using Barker and Pawlewicz's (1986) equation.

Location	Rock Unit	Vitrinite reflectance (R_o)		Thermal alteration index (TAI)/ equiv. R_o (%)	*Calculated paleo-temperature (in °C)
		no. of readings	Mean R_o (%)		
2	Onondaga Fm. (Ls)	---	---	2.3 / 0.7	108
3	Marcellus Fm. (Sh) Onondaga Fm.	50 ---	2.07 ---	3.3 / 1.38 3.2 / 1.25	247/195 182
4	Onondaga Fm.	---	---	3.0 / 1.0	154
5	Onondaga Fm.	---	---	2.7 / 0.93	145
8	Onondaga Fm.	---	---	2.5 / 0.8	126
9	Helderberg Gp. (Ls)	---	---	2.7 / 0.93	145
10	Onondaga Fm.	---	---	2.8 / 0.95	148
11	Onondaga Fm.	---	---	2.9 / 0.98	152
14	Onondaga Fm.	---	---	2.3 / 0.7	108
16	Onondaga Fm.	--- ---	--- ---	2.2 / 0.6 2.3 / 0.7	89 108
17	Onondaga Fm.	---	---	2.8 / 0.95	148
18	Onondaga Fm.	---	---	3.3 / 1.38	195
20	Onondaga Fm.	---	---	2.3 / 0.7	108
21	Onondaga Fm.	---	---	3.6 / 1.75	225
22	Marcellus Fm. Onondaga Fm.	46 ---	1.13 ---	3.3 / 1.38 2.3 / 0.7	170/195 108
24	Onondaga Fm.	---	---	2.7 / 0.93	145
25	Marcellus Fm.	29	0.94	2.5 / 0.8	146/126
32	Marecllus Fm.	49	1.56	2.2 / 0.6	210/89
33	Marcellus Fm.	81	1.66	3.2 / 1.25	219/182
34	Marcellus Fm.	55	1.80	3.2 / 1.25	222/182
36	Marcellus Fm.	9	1.27	3.9 / 2.13	184/251
38	Onondaga Fm.	---	---	2.5 / 0.8	126
39	Tully Fm.	---	---	2.4 / 0.7	108

Table 5 (continued)

40	Tully Fm. (Ls)	---	---	3.8 / 2.0	243
41	Rhinestreet Fm. (Sh) Middlesex Fm. (Sh)	48 5	0.85 1.12	2.2 / 0.6 2.2 / 0.6	133/89
42	Rhinestreet Fm.	47	0.88	2.2 / 0.6	137/89
43	Middlesex Fm.	---	---	2.3 / 0.7	108
48	Lockport Fm. (Dol.)	---	---	3.8 / 2.0	243
50	Clinton Gp. (Sh)	—	—	2.5 / 0.8	126
54	Lockport Fm.	---	---	3.8 / 2.0	243
58	Salina Gp. (Sh)	52	0.97	2.8 / 0.95	150/147
59	Medina Gp. (Sh)	46	0.65	2.2 / 0.6	99/89
63	Utica Fm. (Sh)	---	---	2.3 / 0.7	108
64	Utica Fm. Black River Gp. (Ls)	--- ---	--- ---	3.8 / 2.0 3.8 / 2.0	243 243
65	Black River Gp.	—	—	2.5 / 0.8	126
66	Beekmantown Gp. (Dol.)	---	---	3.6 / 1.75	225
68	Beekmantown Gp.	---	---	3.8 / 2.0	243
70	Onondaga Fm.	---	---	2.2 / 0.6	89
79	Utica Fm.	---	---	3.5 / 1.63	216
80	Utica Fm.	---	---	3.0 / 1.0	154
82	Utica Fm.	---	---	2.6 / 0.90	140
86	Utica Fm.	---	---	2.7 / 0.93	145
93	Utica Fm.	--- ---	--- ---	2.4 / 0.75 2.4 / 0.75	117 117

★ temperatures are calculated from mean Ro or TAI-equivalent mean Ro. Where both Ro and TAI measurements were taken, the first temperature is from mean Ro and the second temperature is from TAI-equivalent mean Ro.

5.4 Stable isotopes

The $\delta^{18}O$ values of the studied cements range from - 4.7 to -13.8‰ PDB and the $\delta^{13}C$ from -4 to +4.5‰ PDB (fig. 18, table 6). There is no clear difference in isotopic composition between calcite and dolomite cements, between cements of different rock units, and there appears to be no geographical trends.

Although negative $\delta^{18}O$ seems to be a general characteristic of most Paleozoic and older carbonates, burial cements are nearly always more depleted in ^{18}O than marine

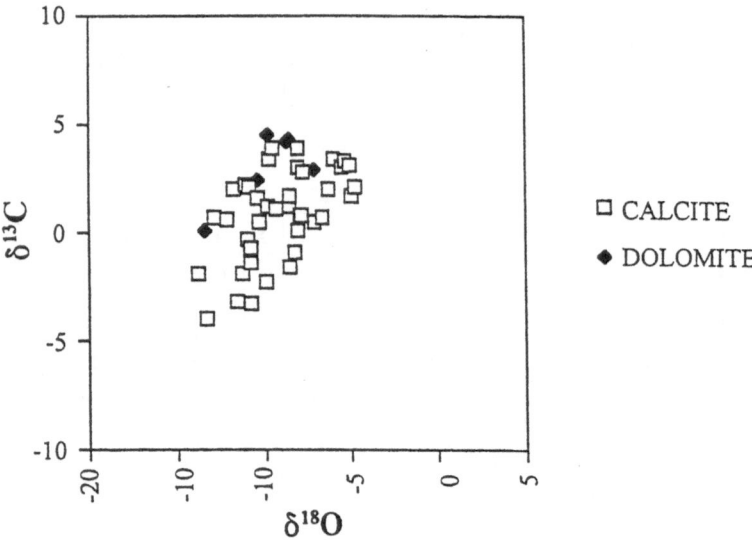

Figure 18: δ¹⁸O vs. δ¹³C diagram for bulk calcite and dolomite cement in the studied samples. See text for discussion.

and early meteoric cements of the same carbonates (Choquette and James, 1987), and heavily depleted δ¹⁸O as well as cross-plotted δ¹³C values of the studied samples (fig. 18) are typical of cements interpreted as of burial origin (Moore, 1985; Lee and Friedman, 1987; Hurley and Lohman, 1989; Barnaby and Reed, 1992; Lavoi and Bourque, 1993; Lee and Bethke, 1994; Qing and Mountjoy, 1994 and many others). A late burial origin of the cement samples is also indicated by the fact that most of the samples came from fractures and vugs that cut across all depositional and almost all diagenetic fabrics of the rocks, except for some stylolites. However, the question is whether the light oxygen isotopic values were acquired from exchange with isotopically light meteoric waters during diagenesis or from high crystallization temperatures under deep burial. For our samples we infer the latter was true. The evidence comes from ice melting temperatures (T_m) which indicate salinities of 5 to 30 wt% NaCl for the same samples from which the isotopic measurements were taken (table 6). Such high salinities point against meteoric waters and toward subsurface brines. The high homogenization temperatures (T_h) measured in these samples (table 6) also provide independent evidence of deep burial.

The δ¹³C values of the samples cover the range commonly observed in ancient as well as Holocene carbonates (Land, 1980, fig. 2). The range also fits very well with the -4 to +4‰ (PDB) of seawater (Lee and Bethke, 1994). Carbon isotopic composition of carbonate cements is a poor indicator of diagenetic conditions. Lack of change in carbon isotopic signature in diagenetic mineral phases from initial seawater values is common and is perhaps due to relative insolubility of CO_2 (Land, 1983).

Table 6: Summary of results of stable isotope study. High fluid salinity and high fluid homogenization temperatures of the samples are indicative of deep brines.

Location	Rock unit	Cement type ◇	$\delta^{13}C$ % PDB	$\delta^{18}O$ % PDB	Mean Th (°C)	*Fluid salinity (wt% NaCl)	*Calculated $^{18}O_{water}$ (SMOW)
1	Onondaga Fm.	Calcite (F)	0.8	-8.0	172.5	19.8	11.2
2	Onondaga Fm.	Calcite (F)	0.7	-6.6	171.9	15.4, 18.8	12.8
3	Onondaga Fm.	Calcite (F)	-0.7	-10.8	151.9	6.7, 5.3	7.2
5	Onondaga Fm.	Calcite (V)	2.8 3.4	-7.8 -4.0	86.7	7.3	4.2 6.1
6	Onondaga Fm.	Calcite (V)	3.0	-5.9	100.1	19.6, 17.9	8.1
8	Onondaga Fm. Helderberg Gp.	Calcite (F) Calcite (V)	1.2 3.3	-8.6 -5.3	112.8 123.1	17.0, 16.5	5.5 10.6
9	Onondaga Fm. Heldeberg Gp.	Calcite (V) Dolomite (V)	2.1 2.9	-4.7 -7.1	138.7 119.3	25.3	12.5 6.0
10	Onondaga Fm.	Dolomite (V)	2.4	-10.4	132.6	3.9	3.9
17	Onondaga Fm.	Calcite (V)	0.6	-12.2	104.3	13.5	1.8
18	Onondaga Fm.	Calcite (V)	1.2	-9.8	98.3	14.3	3.4
20	Onondaga Fm.	Calcite (V)	1.1	-9.4	90.8	---	3.1
22	Tully Fm. Marcellus Fm. Onondaga Fm. Helderberg Gp.	Calcite (F) Calcite (V) Calcite (F) Calcite (F) Calcite (F)	3.9 3.4 1.6 3.4 1.7	-8.1 -9.8 -10.4 -9.8 -8.6	--- 149 43.1 124.5	25.3	--- 8.1 4.83 7.3
23	Onondaga Fm.	Calcite (F)	-3.3	-10.8	---	---	---
24	Onondaga Fm.	Calcite (I)	1.7	-4.9	118.9	20.9	10.6
26	Helderberg Gp.	Calcite (F) Calcite (F)	0.1 0.5	-8.1 -10.3	162.9	---	10.7 8.5

Table 6 (continued)

27	Helderberg Gp.	Calcite (F)	-0.9	-8.30	156.5	5.1, 16.0	10.1
28	Helderberg Gp.	Calcite (V)	2.2	-11.1	126.6	---	4.9
29	Helderberg Gp.	Calcite (V) Calcite (V) Calcite (I)	3.1 2.0 3.0	-5.0 -6.3 -8.1	125.6	10.9, 11.9 13.6, 17.7	11.1 9.8 7.9
36	Marcellus Fm.	Calcite (F)	-3.2	-11.6	---	---	---
40	Tully Fm.	Calcite (F)	2.0	-11.8	---	---	---
45	Lockport Fm.	Dolomite (V)	4.3	-8.6	123	26.9,30.5,31.2	4.2
47	Lockport Fm.	Dolomite (V)	4.5	-9.8	122.9	---	3.6
48	Lockport Fm.	Dolomite (V)	4.2	-8.8	117.6	---	4.1
61	Trenton Gp.	Calcite (F) Calcite (V)	-1.9 -1.4	-11.3 -10.8	161.8	24.5, 25, 25.8, 24.9	7.4 7.9
63	Beekmantown Gp.	Calcite (V)	-1.6	-8.6	112.3	21.2	3.3
64	Utica Fm. Trenton Gp. Black River Gp.	Calcite (F) Calcite (F) Calcite (F)	0.7 2.1 3.9	-12.2 -10.9 -9.6	--- 176.3	15.3, 18.8	--- 8.7 10.0
65	Black River Gp.	Calcite (V) Calcite (F)	0.8 0.5	-7.9 -7.1 -9.4	161	4.1, 8.9, 4.0	10.8 10.9 9.3
66	Beekmantown Gp.	Calcite (F) Calcite (F)	-4.0 -2.3	-13.4 -9.9	138.4	26.6, 24, 23, 23.8, 23.4	3.6 7.1
68	Beekmantown Gp.	Dolomite (V) Calcite (F)	0.1 -1.9	-13.4 -13.8	176.5	24	4.5 5.7

◇ F= fracture-filling, V = vug-filling, I = interparticle

♦ Derived from Tm (see section on "fluid inclusion study")

* $\delta^{18}O_{water}$ was calculated using the following relationships:

for calcite, $10^3 \ln\alpha = 2.78 \times 10^6 T^{-2}$ (°K) - 2.89 (from Friedman and O'Neil, 1977)

for dolomite, $10^3 \ln\alpha = 3.2 \times 10^6 t^{-2}$ (°K) - 3.3 (from Fritz and Smith, 1970)

We have used the mean T_h of the samples to get a rough estimate of the $\delta^{18}O$ of the cement-precipitating waters and the values range between 1.7 to 12.8‰ SMOW (table 6). This is well within the range of modern subsurface brines (Hitchon and Friedman, 1969).

6 COMPARISON AND REFINEMENT OF PALEOTEMPERATURE DATA

In figure 19 paleotemperatures derived from Th_{max}, mean Ro, TAI, and I.C. are plotted against sampling locations. More detailed comparison is shown in table 7. Comparison of all four temperatures at every location was not possible because of sample lithology: generally, carbonate samples yielded Th_{max}, and TAI; Devonian shale samples gave I.C., Ro, and TAI values; older shales gave I.C. and TAI. Moreover, in some dark micritic carbonates only TAI could be measured, and at other locations of carbonate rocks only Th_{max} was measured. Consequently, in several sample locations only one paleotemperature estimate is available.

It is seen that temperatures calculated from I.C. are the highest for the shale locations. TAI measured in the same samples give consistently lower temperatures - often more than 100° C lower - than I.C. Poor correlation between the two are shown in figure 20. The problems of TAI measurement and interpretation have been discussed above. The samples consisted of widely varying percentages of amorphous-sapropelic, herbaceous (spore-pollen), degraded herbaceous, inertinitic, and woody-structured organic matters. These components are known to respond differently to thermal maturation (Tissot et al. 1980) although no correlation was found between the relative amounts of these components and TAI in the studied samples.

In Devonian shales in which vitrinite reflectance analyses were made, temperatures calculated from mean Ro (%) generally fall in between I.C. and TAI (fig. 19) and the correlation between either TAI- and Mean Ro- or I.C.- and Mean Ro-derived temperatures are poor (fig. 20). A somewhat better correlation between I.C. and Ro temperatures is obtained if a measure of the maximum Ro, such as the average of the top 10% of the readings (here called Ro_{max}; see Clayton, 1989, for example) is used (fig. 20) instead of mean Ro. The use of Ro_{max} is perhaps a valid approach for Devonian rocks in which the amount of older recycled vitrinites is minimal.

Th_{max} temperatures could be matched against I.C. temperatures only at four locations (numbers 3, 22, 63, and 64) from which both carbonate and shale samples were collected. In each of these locations Th_{max} temperature is lower than I. C. temperature, about 100° C lower for locations 3 and 22 (fig. 19, table 7). As previously explained, our T_h were not corrected for pressure and therefore Th_{max} of any sample is believed to represent a temperature lower than the maximum burial temperature. Only in three locations (numbers 3, 22, 58) Th_{max} could be directly matched against Mean Ro temperatures (fig. 19). The match is excellent; however, more data are needed to establish a definite relationship. In twenty-two sample

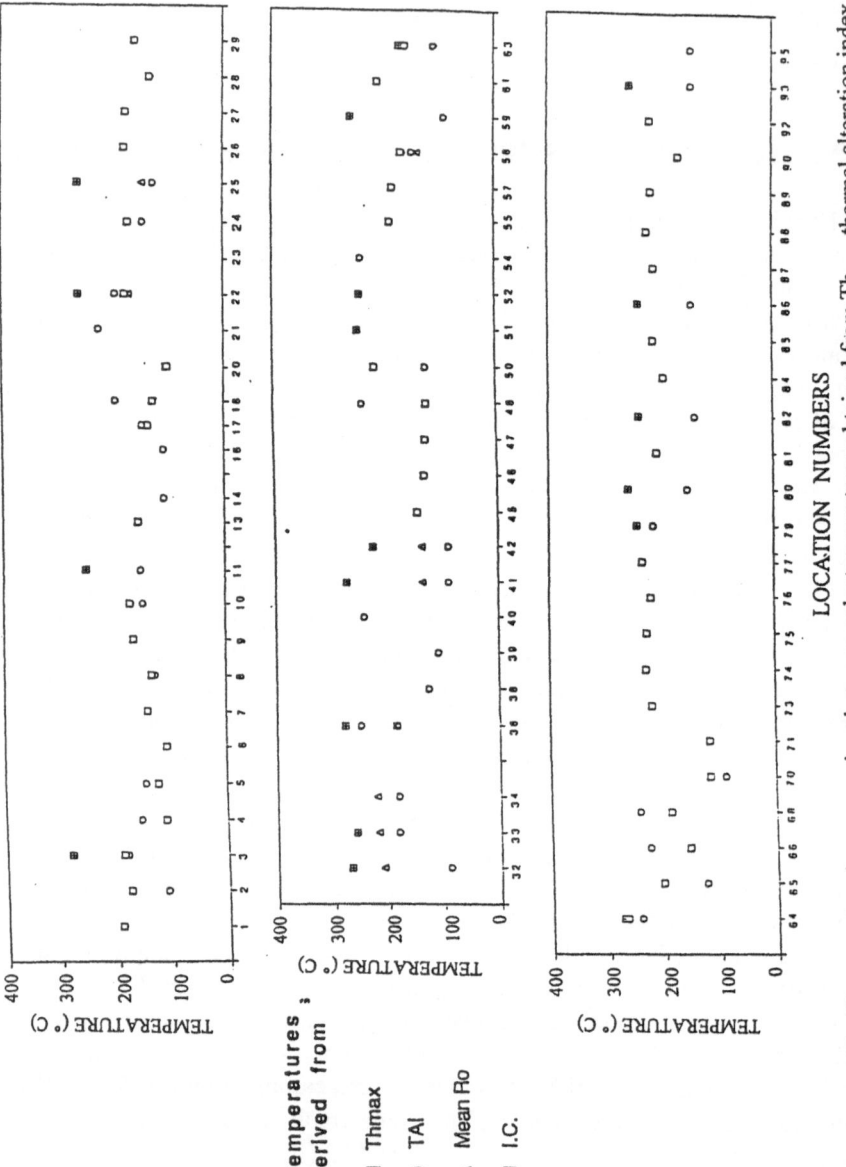

LOCATION NUMBERS

Figure 19: Diagram showing comparison between paleotemperatures obtained from Th_{max}, thermal alteration index (TAI), Mean Ro, and illite crystallinity (I.C.) at different locations. Because of limitations imposed by sample lithology, all four temperatures were not available in most of the locations.

Table 7: Summary table for paleotemperatures at different locations obtained from various techniques.

Location no.	Rock unit	Paleotemperatures (°C) determined from				
		Th_{max}	Mean Ro	TAI	I.C.	% Illite
1	Onondaga Fm.	193	---	---	---	---
2	Onondaga Fm.	176.9	---	108	---	---
3	Onondaga Fm.	188.7	---	---	---	---
	Marcellus Fm.	---	247	182	280	> 200
4	Onondaga Fm.	108.7	---	154	---	---
5	Onondaga Fm.	122	---	145	---	---
6	Onondaga Fm.	106	---	---	---	---
7	Onondaga Fm.	112	---	---	---	---
	Helderberg Gp.	140	---	---	---	---
8	Onondaga Fm.	132.4	---	---	---	---
	Helderberg Gp.	131	---	126	---	---
9	Onondaga Fm.	166	---	---	---	---
	Helderberg Gp.	157.6	---	---	---	---
10	Onondaga Fm.	172	---	148	---	---
11	Marcellus Fm.	---	---	---	255	> 200
13	Onondaga Fm.	156.3	---	---	---	---
14	Onondaga Fm.	---	---	108	---	---
16	Onondaga Fm.	---	---	89, 108	---	---
17	Onondaga Fm.	136.6	---	148	---	---
18	Onondaga Fm.	127.5	---	195	---	---
20	Onondaga Fm.	100.4	---	108	---	---
21	Onondaga Fm.	---	---	225	---	---
22	Geneseo Fm.	---	---	---	255	> 200
	Marcellus Fm.	178	170	---	260	> 200
	Onondaga Fm.	71.3	---	108	---	---
	Helderberg Gp.	169.7	---	---	---	---
24	Onondaga Fm.	171.5	---	145	---	---
25	Marcellus Fm.	---	146		260	> 200
26	Helderberg Gp.	178	---	---	---	---
27	Helderberg Gp.	175.6	---	---	---	---
28	Helderberg Gp.	132	---	---	---	---
29	Helderberg Gp.	159.3	---	---	---	---
32	Marcellus Fm.	---	210	---	275	> 200
33	Marcellus Fm.	---	219	---	260	> 200
34	Marcellus Fm.	---	222	---	---	---
36	Marcellus Fm.	---	182	---	280	> 200
38	Onondaga Fm.	---	---	126	---	---
39	Tully Fm.	---	---	108	---	---
41	Dunkirk Fm.	---	---	---	260	> 200
	Rhinestreet Fm.	---	---	89	275	> 200
42	Rhinestreet Fm.	---	---	---	225	> 200
43	Middlesex Fm.	---	---	108	---	---
45	Lockport Fm.	143.8	---		---	---
46	Lockport Fm.	130.4	---	---	---	---
47	Lockport Fm.	127.8	---	---	---	---
48	Lockport Fm.	125	---	---	---	---
49	Lockport Fm.	140.3	---	---	---	---
50	Clinton Gp.	218	---	---	---	---
51	Salina Gp.	---	---	---	255	> 200
52	Salina Gp.	---	---	---	245	> 200
55	Clinton Gp.	188	---	---	---	---
57	Salina Gp.	182.6	---	---	---	---
58	Salina Gp.	166.8	150	---	---	---

Table 7 (continued)

59	Medina Gp.	—	99	89	225	> 200
61	Black River Gp	210	—	—	—	—
63	Utica Fm.	—	—	147	246	> 200
	Beekmantown Gp.	159.3	—	—	—	—
64	Utica Fm.	—	—	242	275	> 200
	Black River Gp.	269	—	—	—	—
65	Black River Gp.	203.6	—	—	—	—
66	Beekmantown Gp.	155.3	—	—	—	—
68	Beekmantown Gp.	187.2	—	—	—	—
70	Onondaga Fm.	116.6	—	—	—	—
71	Onondaga Fm.	118	—	—	—	—
73	Beekmantown Gp.	219.7	—	—	—	—
74	Beekmantown Gp.	230.2	—	—	—	—
75	Beekmantown Gp.	228.7	—	—	—	—
76	Beekmantown Gp.	220	—	—	—	—
77	Beekmantown Gp.	236	—	—	—	—
79	Utica Fm.	—	—	216	246	> 200
80	Utica Fm.	—	—	154	260	> 200
81	Beekmantown Gp.	208.5	—	—	—	—
82	Utica Fm.	—	—	140	240	> 200
84	Trenton Gp.	195.2	—	—	—	—
85	Black River Gp.	213.5	—	—	—	—
86	Utica Fm.	—	—	145	240	> 200
87	Black River Gp.	211.3	—	—	—	—
88	Black River Gp.	223.8	—	—	—	—
89	Black River Gp.	216	—	—	—	—
90	Beekmantown Gp.	167	—	—	—	—
92	Beekmantown Gp.	218	—	—	—	—
93	Marcellus Fm.	—	—	145	—	—
95	Marcellus Fm.	—	—	145	—	—

locations, Th_{max} of carbonate rocks could be directly compared to TAI of the same samples, but the correlation was found to be very poor (fig. 20).

It is difficult to ascertain from the data which, if any, of these variously determined paleotemperatures best represents the true maximum paleotemperatures experienced by the rocks. However, the data allows us to estimate a range of maximum paleotemperatures. The upper limits of the maximum paleotemperatures are clearly represented by I.C. temperatures. The validity of I.C. temperatures is strengthened by a reasonably close match with Ro_{max} temperatures (fig. 20). We believe that correlation between the two will only increase if larger numbers of vitrinite reflectance readings are taken from the samples. However, I.C. measurements were made from only 19 of our samples; therefore, another paleotemperature measurement has to be applied to the rest of the locations. Mean Ro or Ro_{max} temperatures can not be used because Ro measurements were made only in Devonian shales from ten locations all of which were also included in I.C. analysis. The TAI data is inconsistent with all others and has proved least useful for determining paleotemperatures in this study. Th_{max} is a more suitable measure of 'maximum' paleotemperature for the rest of the locations not only because it is a more direct method of determining paleotemperature, but also because most of our data comes

71

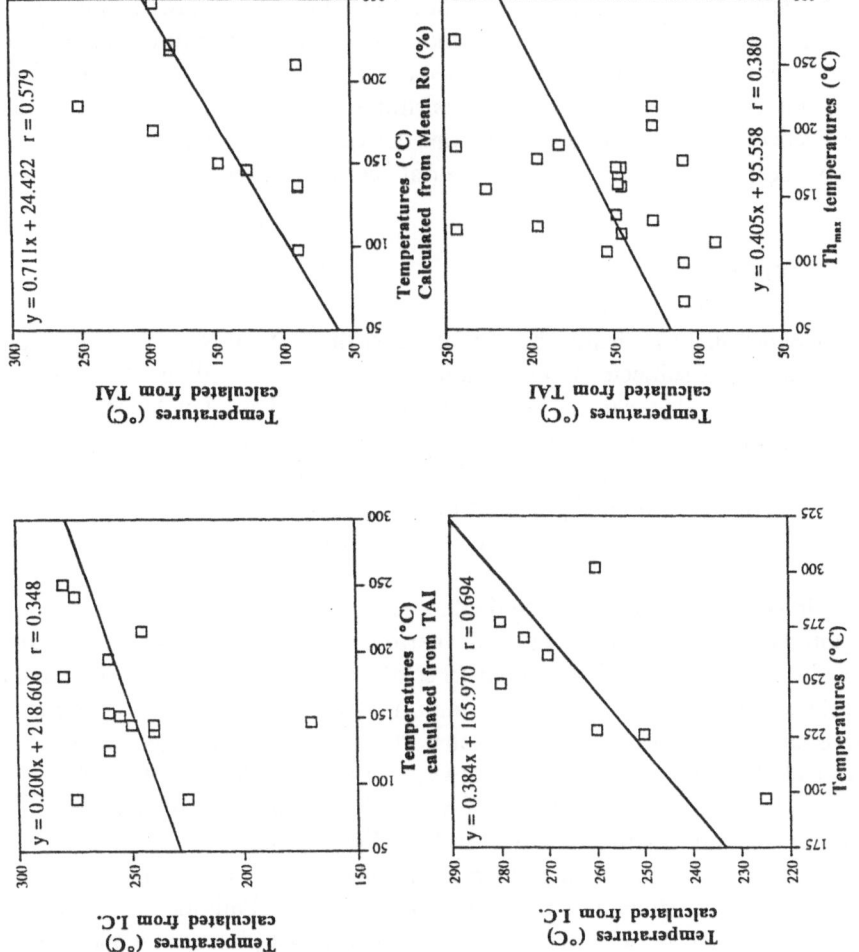

Figure 20: Correlation diagrams between temperatures obtained from different techniques.

from T_h measurements - from a total of 53 locations. However, as previously discussed, Th_{max} can not be considered the maximum paleotemperatures in the same sense of I.C. temperatures. It is to be considered a lower limit of the maximum paleotemperature and it serves as a useful constraint in absence of I.C. temperatures. Thus, two sets of paleotemperatures data - one derived from I.C. and the other from Th_{max} - have been used for calculations of burial depths (section 8).

7 SOURCE OF HEAT

Paleotemperature analysis clearly shows that across the State of New York rocks at or near the surface were subjected to high temperatures, often exceeding 200°C. Included in this analysis are many Middle Devonian as well as a few Upper Devonian rock samples. If there was indeed very little sedimentation in New York after the Devonian, it becomes tempting to interpret these high paleotemperature signatures as due to elevated paleogeothermal gradients (see Tillman and Barnes, 1983, for example). In the following section, we present arguments against high paleogeothermal gradients.

Increase in geothermal heatflow and the thermal gradient may occur locally due to igneous intrusions. However, except for the narrow belt of Jurassic ultramafic dike intrusions in central New York, scattered dikes of Cretaceous age in eastern parts of the Adirondacks (McHone, 1978; pers. comm., 1994), small Devonian and younger (?) plutons and ultramafic dikes in the southeastern New York - Connecticut border area (Mose et al. 1976, Miller and Kent, 1989), and the early Jurassic Palisade diabase sills (Dunning and Hodych, 1990) west of New York City, there is no evidence of Phanerozoic igneous activity in New York State.

The Palisade sills are more than 100 km southeast of the nearest sampling locations (1, 2, 36, fig. 9) of this study. It is therefore unlikely that the Palisade event had any reheating effect on our samples. The Devonian and younger plutons and dikes of southeastern New York - southwestern Connecticut area may likewise be eliminated from possible causes of igneous reheating of our samples. According to Mose et al. (1976) one of the plutons in this area, the Peekskill Granite, and associated dikes were emplaced at shallow depth, and "contact metamorphic effects are commonly absent because the granite was not of sufficiently high temperature or cooled too quickly". The closest igneous body from this group of plutons and dikes to our sample locations 1, 2, 3, and 26 (fig. 9) is more than 60 km away.

Although we hold the view that all of the above-mentioned igneous activities were too small to cause any significant reheating of the studied rocks, the Jurassic dike swarms of central New York perhaps deserve a greater attention, because these dikes have been mentioned as a possible cause of locally elevated paleotemperatures (Karig, 1987).

Most dikes act as geologically instantaneous heat pulses that can heat up the host rock only around their immediate contacts (Delaney, 1987). McCelland-Brown

(1981) shows that for a 1.25m thick dike in Scotland, temperature decreased exponentially in the host rock, already having an ambient temperature of 150° C, from 760° C at the contact to 375° C only at a distance of 1.2m from the contact. Buchan and Schwarz (1987), in their study of remanent magnetization in the host rock around a 38m thick dike of the Northwestern Territories, found that at a distance of 37m from the contact the remanent magnetization remained undisturbed, indicating that heat, even from such a thick dike, had failed to significantly raise the temperature (and reset the remanent magnetization) of rocks only at a distance of 37m.

Among the 82 dikes reported from a NW-SE trending narrow zone - 25km by 45km - in central New York, the thickest one is only 3.5m thick (Kay et al. 1983; Kay and Foster, 1986). Another dike reaches a thickness of 1m, and the rest are between a few cm and 0.75m thick. Heat generated from these thin dikes could not have possibly traveled very far. Moreover, the dike swarm was emplaced over an interval of 145±7 to 120 Ma (Kay et al. 1983), implying that there was no possibility of a combined heat effect. Evidence, such as sharp contacts of these dikes with the host rock, lack of metamorphism in the adjacent rocks, texture, and vertical extent of very thin dikes that are believed to have originated in the upper mantle have led workers to believe that these dikes were not emplaced by magmatic flows, but by "fluidization mechanism", in which fragmented materials were transported in a fast- moving gas stream (Foster, 1970; Reitan et al. 1970; Kay et al. 1983; Kay and Foster, 1986). If this interpretation is correct, the possibilities of overheating the surrounding rocks and establishing a sustained, elevated geothermal gradient seem even more remote, because a gaseous flow has a much lower heat capacity than a magmatic flow. Moreover, because most dikes cool very quickly, hydrothermal circulation, which is known to be an important process during cooling of large igneous bodies, fail to become established (Delaney, 1987).

Considering these arguments, it is inferred the late Jurassic dikes had played little or no role in overprinting the paleotemperature signatures of rocks of central New York, except possibly at their immediate contacts. Although many of our samples are from the general area underlain by these dikes, none of these samples are from within several kilometers of the known occurrence of the dikes. Much of the arguments presented above also applies to the early Cretaceous dikes of east-central Adirondacks. The known occurrences of these mafic dikes (McHone, 1978, fig. 1) are more than 80 km north of our nearest sample locations (73, 74, 75; fig. 9) and it is highly unlikely that the heating effect of these dikes extended to our study area.

Another commonly invoked mechanism for elevated geothermal gradient is rooted in many models of MVT mineralization. MVT minerals showing high fluid-homogenization temperatures but lacking direct evidence of former deep burial have often been interpreted as having precipitated from compaction-driven, warm, deep basinal fluids that have migrated to shallow (and cooler) depths, especially to basin margins (Sharp, 1978; Cathles and Smith, 1983). Tectonism at the basin margin (orogenic belt) has also been proposed as a mechanism for releasing large amounts of fluids and heat causing craton-ward fluid/heat migration across foreland basins (Oliver, 1986; Deming, 1992). The "gravity driven" flow model (Garven and Freeze, 1984) calls for the topographically high orogenic belt to serve as the recharge area for cool meteoric water which would subsequently sink into the deeper part of the basin, become hotter, and emerge at the craton margin (see also Deming and Nunn, 1991). A combination of these models have been used to interpret spatial variations in ancient and present geothermal gradients within sedimentary basin (Hitchon, 1984; Garven, 1989; Tilley et al. 1989; Deming et al., 1992; Qing and Mountjoy, 1992).

An important implication of these models is that minerals forming from such basinal fluids do not reflect the normal burial temperatures of an area and, therefore, paleotemperature data (T_h, for example) obtained from them may not be used directly to estimate their former depths of burial on the basis of a presumed "normal" geothermal gradient.

Examination of the spatial distribution of Th_{max} paleotemperatures determined in this study (fig. 21), however, fails to reveal any trend that can be attributed to cooling of deep basinal fluid at shallow depths. For the "gravity driven" flow model to be applicable, lower paleotemperatures should be found along the eastern part of the study area which was in proximity to topographically high areas at least since the Taconic orogeny and higher paleotemperatures should be found in western New York and beyond. The opposite is true for our overall paleotemperature data as well as for individual rock units like the Onondaga and Beekmantown (fig. 21).

Our data would better fit the "tectonically driven" model (Oliver, 1986; Deming, 1992) except that in central New York almost all paleotemperatures are higher than in areas immediately to the east and west (especially noticeable in the Onondaga Formation): a uniform spatial thermal gradient from the tectonically active area of the east to the craton of the west is not observed. Moreover, the Th_{max} of the Beekmantown, Black River and Trenton groups do not show any systematic spatial variation at all. For example, the Th_{max} of the Beekmantown Group of locations 73 and 76 on the eastern margin of the Adirondacks and closer to the eastern orogenic belt are essentially the same as that in location 92 on the western margin of the Adirondacks across a distance of 200km (fig. 21). Also, vertical, up-formational flow of hot fluid to shallow depths can be best disproved by considering paleotemperatures of the Beekmantown and Black-River groups. The Beekmantown Group is the basal sedimentary rock unit in the northern Appalachian basin. The Black River Group is also the basal unit between location 61 and 89 (fig. 21) from where the Beekmantown strata were removed by erosion and the Black River limestones sit unconformably on the Grenville basement. It is very unlikely that hot fluids migrated upward from the depths of the Grenvillian metamorphic basement - a very poor aquifer at best - and precipitated cements from which the T_h measurements have been taken.

There is also no significant difference in T_h of fracture-filling cements and interparticle cements in these rocks or in the younger Devonian carbonates elsewhere. Since fractures are believed to be common conduits for upward migration of fluids, cements precipitated from them is expected to have higher T_h which has not been

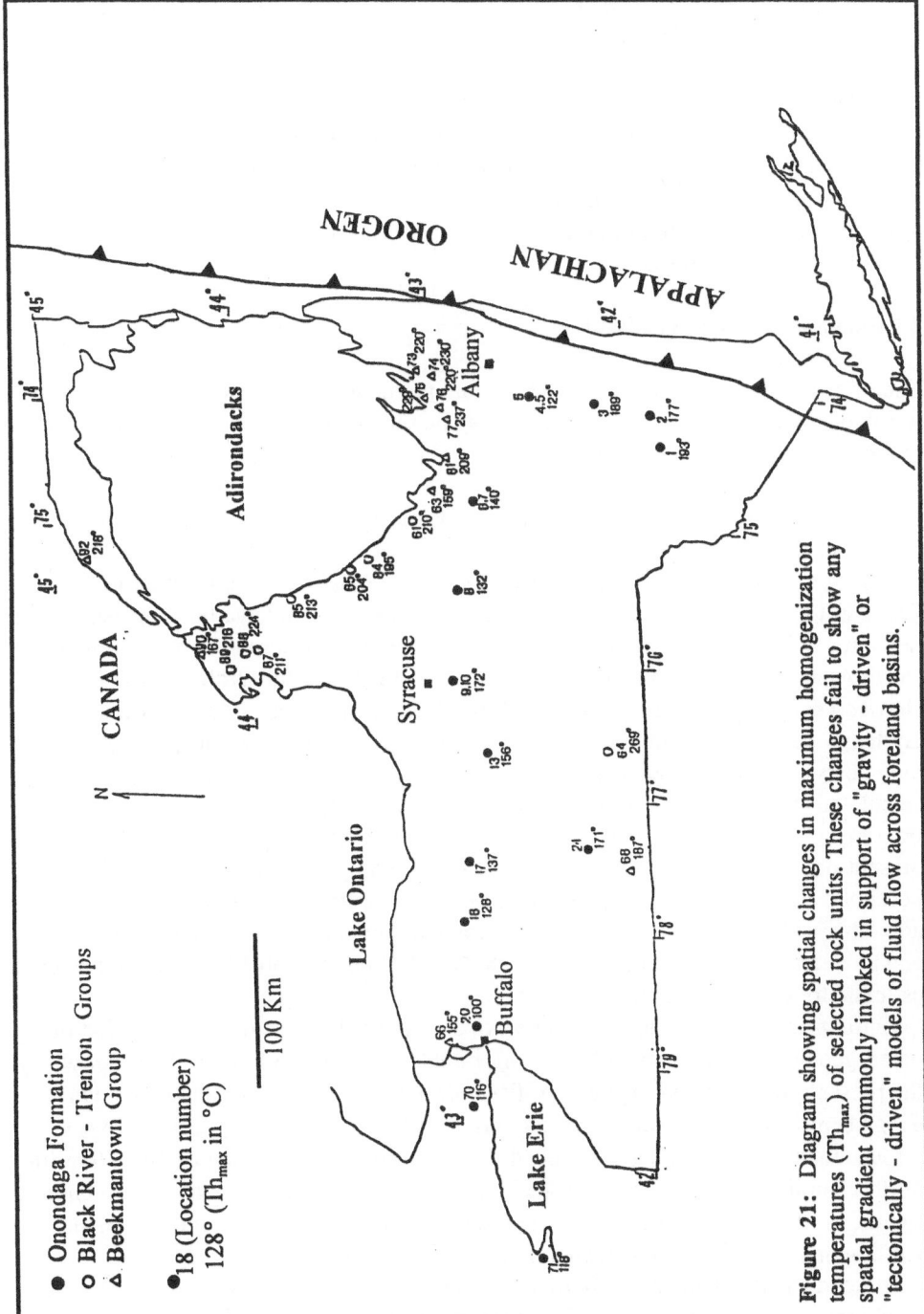

Figure 21: Diagram showing spatial changes in maximum homogenization temperatures (Th$_{max}$) of selected rock units. These changes fail to show any spatial gradient commonly invoked in support of "gravity - driven" or "tectonically - driven" models of fluid flow across foreland basins.

observed in these rocks. Furthermore, the relatively impermeable cover provided by the Late Ordovician Utica Shale and the Middle Devonian Marcellus Shale in New York probably greatly impeded any up-formational flow.

The lack of any systematic spatial trend in New York State is seen not only in T_h values but also in I.C. and Ro. We believe that these paleotemperature signatures in the studied rocks of New York were attained primarily from normal geothermal heat as a function of burial depth which varied spatially across the state. Even if episodes of large-scale basinal fluid migration took place, they probably failed to perturb the normal geothermal regime.

7.1 Paleo-geothermal gradient

Use of a geologically viable paleogeothermal gradient is critical for any burial study. A paleogeothermal gradient for an area may be reconstructed from measured changes of a paleotemperature indicator through a given stratigraphic interval in the area (Levine, 1986; Stern and Reesman, 1986; Lacazette, 1991). In absence of such information, one has to rely on the tectonic history of the basin and perhaps use data from modern analogs.

In most modern sedimentary basins temperature gradients range from 20 to 45° C/km, but in recent rift basins gradients as high as 80° C/km have been measured (Hanor, 1987). Geothermal gradients measured today in eastern and mid-continent North America ranges from 20° C/km to 40° C/km (Jones and Wallace, 1974; Harrison et al. 1983; Hodge, 1984). In New York, the present day geothermal gradients, calculated from bottom- hole temperatures of relatively deep boreholes, range between 20° C/km and 36° C/km, although locally higher gradients have been calculated and generally attributed to convective upwelling through subsurface fracture system (Hodge, 1984). In eastern New York, no information on the present-day geothermal gradients is available, probably because of paucity of deep boreholes. However, relatively low heat-flow measurements from the Adirondacks and other areas of eastern New York are comparable to those in western New York, and an average geothermal gradient of about 30° C/km is probably a reasonable estimate for the eastern half of the state as well.

Findings in New York and other areas of similar geologic setting suggest that the paleo-geothermal gradients of New York might have been within the range shown by the present gradients. On the basis of fission-track dates of the Middle Devonian Tioga bentonite and the bentonite bed within the Upper Ordovician Black River Group, Johnsson (1985) had calculated a paleotemperature gradient of 20° C/km for north-central New York. From vertical vitrinite reflectance profiles of two boreholes, Levine (1986) calculated a paleotemperature gradient of 33° C/km for the southern Anthracite District of Pennsylvania. In central Pennsylvania, Lacazette (1991) calculated an Alleghanian geothermal gradient of 31° C/km from T_h of methane-brine fluid inclusions in quartz veins of the Ordovician Bald Eagle sandstone. In the Allegheny Plateau of western Pennsylvania, thermal modeling by Zhang and Davis

(1993) shows that the geothermal gradient during the Alleghanian orogeny was between 26° C/km and 33° C/km. In the Alberta Basin of Canada, estimated paleotemperature gradients range from 27° C/km to 40° C/km in which the gradients are believed to have increased from the Rocky Mountains in the west to the plains in the east (Hitchon, 1984; Connolly, 1989; Tilley et al. 1989).

Therefore, in this study an average paleo-geothermal gradient of 30° C/km was used for the entire sampling area in New York State. Previous studies in New York that used a gradient of 25-30° C/km include those of Friedman and Sanders (1982), Lakatos and Miller (1983), Johnsson (1985, 1986), Friedman (1987a) and Gerlach (1987). McKenzie (1981) and Issler (1984) recommend the use of present-day geothermal gradient in predicting and interpreting thermal maturation data in thermally old (more than 150 m.y.) basins.

8 BURIAL DEPTHS AND THICKNESSES OF POST-DEVONIAN STRATA

On the basis of a paleogeothermal gradient of 30° C/km and a mean annual surface temperature of 20° C, the maximum burial depths of the studied sampled have been calculated (tables 8, 9). The present-day average annual surface temperature of New York is about 12°C (Hodge et al, 1982), but at the close of the Paleozoic, when the maximum burial of the studied rocks most probably took place, this area was closer to the equator and had a higher inferred average surface temperature. The calculations have been made for both Th_{max} and I.C. temperatures. In sample locations where Th_{max} from more than one rock unit were measured (such as locations 7, 8, 9, and 22), only the maximum Th_{max} obtained has been used. In locations 8, 9, and 22, the Th_{max} of stratigraphically higher units are greater than that of lower units and in location 7, the Th_{max} of the Helderberg Group is 28°C greater than that of the overlying Onondaga Formation (table 7) although the stratigraphic separation of the studied samples of these two units is only a few tens of meters. It is assumed that if more fluid inclusions were available, the Th_{max} data would be consistent with stratigraphic distance between samples.

From the two sets of maximum burial depths, the projected thicknesses of younger rocks (up to end-Devonian in age) on the studied samples at different locations have been subtracted to obtain the probable thicknesses of post-Devonian strata that might have been present at one time at these locations. Thickness data for various rock formations in New York State were taken from Kreidler et al. (1972), Rickard (1975), and Fisher (1977). Only for location 92, the thickness of the Ordovician section was projected from the preserved block faulted rocks of the nearby Ottawa region (Poole et al. 1970, chart II). Projections of sections younger than the studied rocks of different location have been made from due south of the locations, where more complete sections of the younger rocks exist. In these areas, however, only incomplete sections of the Middle and Upper Devonian are present (Rickard, 1975, pl. 2). Therefore, the maximum thicknesses of these rocks found farther west first had

Table 8: Table showing the calculated minimum thicknesses of possible post-Devonian strata at different study locations based on Th_{max} data.

Location no.	Rock unit	Th_{max} (° C)	Burial depth (km)	*Projected thickness of younger rocks (km)	Minimum thickness of post-Devonian strata removed (km)
1	Onondaga	193	5.76	3.63	2.13
2	Onondaga	176.9	5.23	3.45	1.78
3	Onondaga	188.7	5.62	3.55	2.07
4,5	Onondaga	122	3.4	3.2	0.2
6,7	Helderberg	140	4.0	3.1	0.9
8,28	Onondaga	132.4	3.74	2.88	0.86
9,10	Onondaga	172	5.06	2.54	2.52
13	Onondaga	156.3	4.54	2.44	2.10
17	Onondaga	136.6	3.88	2.07	1.81
18	Onondaga	127.5	3.58	1.4	2.18
20	Onondaga	100.4	2.66	1.10	1.56
22	Marcellus	178	5.26	2.38	2.88
24	Onondaga	171.5	5.05	2.05	3.0
29	Helderberg	159.3	4.64	2.78	1.86
45	Lockport	143.8	4.12	1.37	2.75
46	Lockport	130.4	3.68	1.34	2.34
47	Lockport	127.8	3.59	1.27	2.32
48	Lockport	125	3.5	1.6	1.9
49	Lockport	140.3	4.0	1.17	2.73
50	Clinton	218	6.6	2.5	4.1
51	Salina	210	6.33	2.28	4.05
55	Clinton	188	5.6	2.37	3.23
57	Salina	182.6	5.42	2.85	2.57
58	Salina	166.8	4.89	2.57	2.32
61	Trenton	210	6.33	3.86	2.47
63	Beekmantown	159.3	4.65	3.85	0.8

Table 8 (continued)

64	Black River	269.3	8.3	3.86	4.44
65	Black River	203.6	6.12	3.78	2.34
66	Beekmantown	155.3	4.51	2.1	2.41
68	Beekmantown	187.2	5.57	4.64	0.93
70	Onondaga	116	3.2	0.95	2.25
71	Onondaga	118	3.26	0.95	2.31
73	Beekmantown	219.7	6.65	5.55	1.1
74	Beekmantown	230.2	7.0	5.6	1.4
75	Beekmantown	228.7	6.95	5.5	1.45
76	Beekmantown	220	6.66	5.4	1.26
77	Beekmantown	236.6	7.22	5.4	1.8
81	Beekmantown	208.5	6.28	4.68	1.6
84	Trenton	195.2	5.84	3.73	2.11
85	Black River	213.5	6.45	3.57	2.88
87	Black River	211.3	6.37	3.5	2.87
88	Black River	223.8	6.79	3.5	3.29
89	Black River	216	6.53	3.5	3.03
90	Beekmantown	167	4.9	3.5	1.4
92	Beekmantown	218	6.6	3.88	2.72
X	Chazy	206	6.2	5.56	0.64
Y	Beekmantown	211	6.36	5.52	0.84

$\star$ = Stratigraphic separation between the sample and the projected top of the Devonian on the sample location.

X = Taken from Brockerhoff and Friedman (1987)

Y = Taken from Urschel and Friedman (1984)

to be extended eastward, added to the underlying rocks, and then projected northward onto the study locations. For reasons explained earlier, the post-Devonian thicknesses estimated from Th_{max} should be taken as a minimum limit and those calculated from I.C. temperatures as a maximum limit of "maximum" thicknesses of post-Devonian strata. Two Th_{max} data points, shown as locations X and Y (table 8, fig. 22), are taken from fluid inclusion measurements of previous studies (Brockerhoff and Friedman 1987; Urschel and Friedman, 1984) in order to facilitate drawing isopach lines through the northeastern margins of the Adirondacks which was not covered by this study.

Figures 22 and 23 show two isopach maps drawn on the basis of the two sets of post-Devonian thicknesses from tables 8 and 9, respectively. Although the maps have been drawn on the basis of two different sets of data and the one is considerably more

Table 9: Table showing the calculated maximum thicknesses of possible post-Devonian strata at different study locations based on illite crystallinity temperatures.

Location no.	Rock unit	I.C. Temp. (° C)	Burial depth (km)	*Projected thickness of younger rocks (km)	Maximum thickness of post-Devonian strata removed (Km)
3	Marcellus	280	8.67	3.0	5.67
11	Marcellus	250	7.67	2.5	5.17
22	Marcellus	260	8.0	2.36	5.62
25	Marcellus	260	8.0	2.03	6.03
32	Marcellus	275	8.5	3.08	5.42
33	Marcellus	260	8.0	3.0	5.0
36	Marcellus	280	8.67	3.0	5.67
39	Geneseo	250	7.67	2.0	5.67
41	Rhinestreet	275	8.5	1.6	6.9
42	Rhinestreet	225	6.8	1.67	5.13
51	Salina	250	7.67	2.2	5.47
52	Salina	245	7.5	2.2	5.3
59	Medina	225	6.83	1.0	5.83
63 ♠	Utica	200	6.3	3.85	2.45
64	Utica	275	8.5	3.7	4.8
79	Utica	245	7.5	4.3	3.2
80	Utica	260	8.0	4.3	3.7
82	Utica	240	7.26	3.7	3.56
86	utica	240	7.26	3.4	3.83
93	Utica	255	7.83	3.5	4.33

★ = Stratigraphic separation between the sample and the projected top of the Devonian on the sample location.

♠ = The temperature is taken from percentage of illite in stead of I.C. since the latter was found to be anomalously low (see table 1).

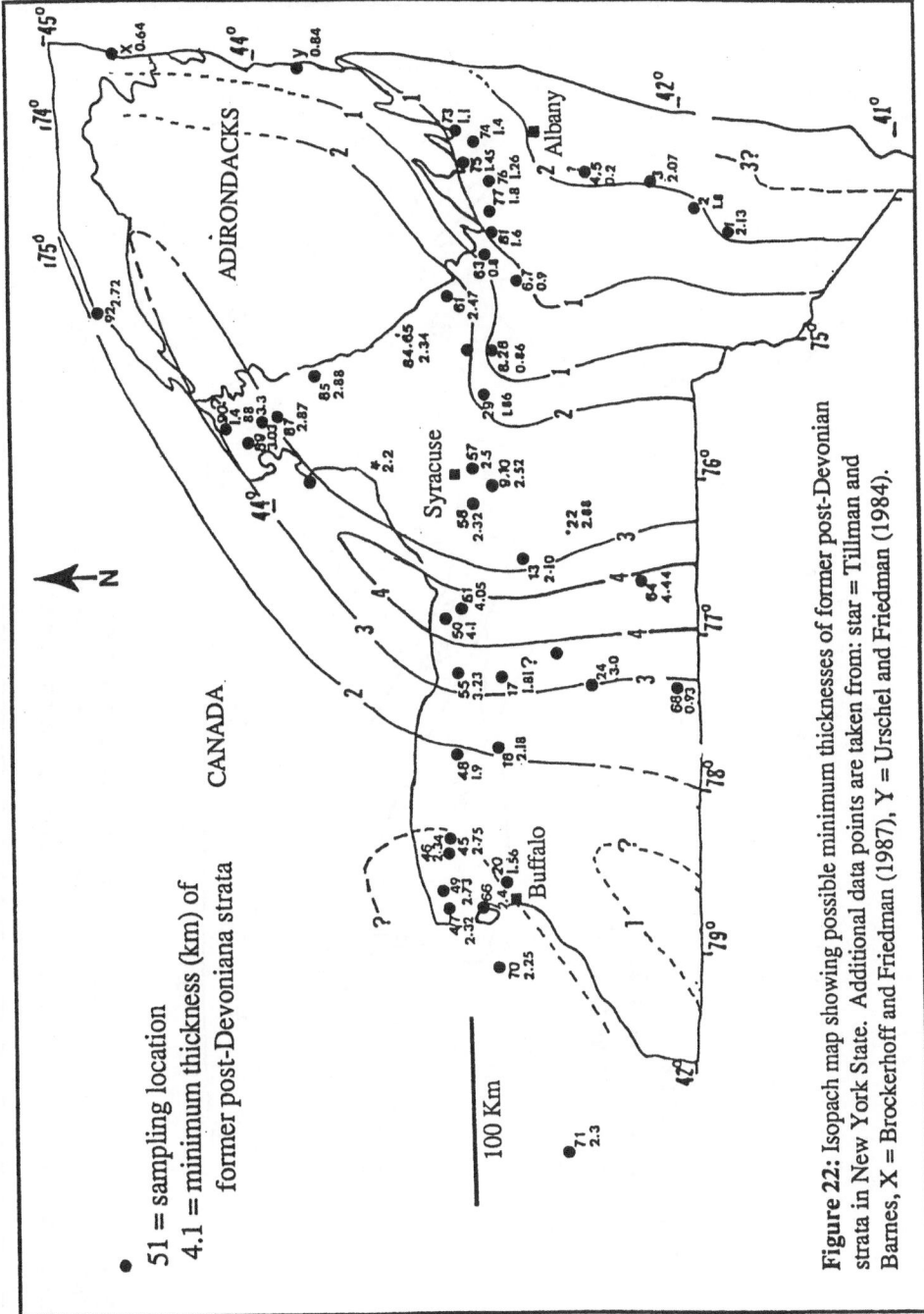

Figure 22: Isopach map showing possible minimum thicknesses of former post-Devonian strata in New York State. Additional data points are taken from: star = Tillman and Barnes, X = Brockerhoff and Friedman (1987), Y = Urschel and Friedman (1984).

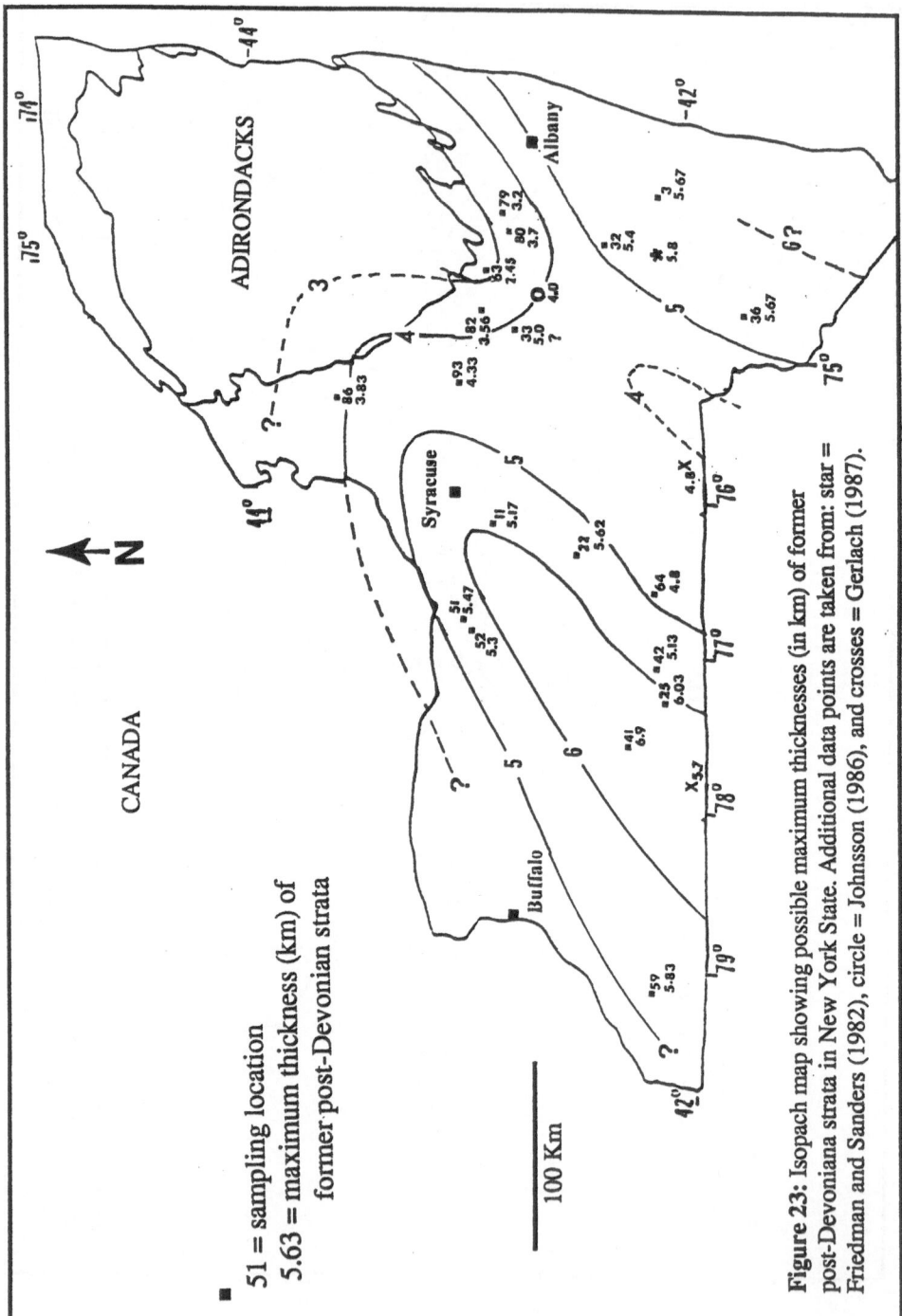

Figure 23: Isopach map showing possible maximum thicknesses (in km) of former post-Devonian strata in New York State. Additional data points are taken from: star = Friedman and Sanders (1982), circle = Johnsson (1986), and crosses = Gerlach (1987).

detailed than the other, there appear to be some basic similarities between them. Both show areas of thicker post-Devonian strata over central and southeastern New York State. Both maps also show a relatively narrow band of thinner post-Devonian strata separating the central and southeastern New York. In the map based on Th_{max} temperatures (fig. 22), there appears to be a third area of thicker post-Devonian strata in westernmost New York which does not show up in the other map because of lack of I.C. data from that area. However, it is possible that at locations 18 and 48 the thicknesses have been underestimated and the 2km isopach line actually passes through areas farther west, beyond location 71 in western Ontario. On the map of maximum post-Devonian strata (fig. 23) thicknesses inferred by other studies have been included in order to compare with the results of this study. The comparisons, if not good, are nevertheless satisfactory.

9 POST-DEVONIAN GEOLOGY OF NEW YORK STATE

If New York State and the adjacent areas were indeed covered by a substantial thickness of post-Devonian strata, many questions about the post-Devonian geologic history of the region have to be considered. These include the duration of post-Devonian sedimentation, the source(s) of sediments, paleogeographic and depositional systems, and tectonics. There are no direct means of interpreting these aspects of the post-Devonian geology now, but it is possible to draw many geologically reasonable inferences.

The nearest occurrence of post-Devonian strata of significant thickness to the study area is in Pennsylvania and, therefore, the possibility of the post-Devonian strata in New York State being an extension of those in Pennsylvania has to be considered. In eastern Pennsylvania, the combined thickness of the Carboniferous rocks is 3.4km (COSUNA Chart, 1985). However, the thickness decreases in practically every direction from eastern Pennsylvania, and the Carboniferous rocks show a dominantly NW-directed sediment dispersal pattern (Pelletier, 1958; Meckel, 1970). Moreover, except for eastern and southeastern Pennsylvania where Carboniferous sedimentation was continuous, an erosional unconformity separates the Mississippian Mauch Chuck Formation from the Pennsylvanian Pottsville Formation in other areas. Along the southern border of New York a maximum thickness of only 32m of the Early Mississippian Knapp Formation (equivalent to Mauch Chuck) is unconformably overlain by a 28m thick Middle Pennsylvanian Olean Conglomerate, which is equivalent of the Pottsville Formation (COSUNA Chart, 1985). The uplift that caused the removal of pre-Olean strata is possibly reflected in the reset $^{40}Ar/^{39}Ar$-age of about 320 MA measured in authigenic feldspar of Lower Ordovician dolostones of New York (Friedman, 1990; 1994).

In eastern Pennsylvania, about 1km thick Upper Pennsylvanian strata (the Llewellyn Formation) is preserved (COSUNA Chart, 1985). Supposing that the equivalents of these strata were once present in New York above the Olean

Conglomerate, it is still difficult to account for much more than 1km of total Carboniferous sediments which might have had their provenance in the southeast. Although the post-Devonian strata in southeastern New York (figs. 22, 23) may have derived part of their sediments from this source, it is unlikely that the rest of New York received much sediment from the same source. It is possible that much of the post-Devonian strata in New York was in fact post-Pennsylvanian in age. Assuming this, what could be the provenance of these sediments?

A likely sediment source is inferred from the interpretation of Levine (1986) who suggests Carboniferous rocks of the anthracite region of eastern and northeastern Pennsylvania were buried under 6-9km thick thrust-sheets (or under sediments derived from them) before Alleghanian folding (ca. 290 - 270Ma) occurred. It is likely that sediments derived from these tectonically emplaced blocks, straddling, or perhaps resting directly on, southeastern New York (Beaumont et al. 1987) found their way over much of the adjacent areas of New York. If sediments accounting for about 1km thick Carboniferous strata were already present in southeastern New York when the proposed thrusting event(s) took place, the remaining 4 - 5 km or thicker strata in these areas might have had their provenance in these thrust blocks.

The southeastern source, however, probably fails to account for the thicker sediment accumulation over west-central New York and beyond (figs. 22, 23). The presence of the NE-oriented zone of low sediment accumulation over eastern New York possibly marked the depositional limit for sediment derived from the southeast. If so, areas west of this zone might have received a sediment supply from somewhere else. One such source area might have been the Findlay - Algonquin Arch northwest of the Adirondacks (figs. 1, 24).

We speculate that, in response to rapid tectonic loading of eastern Pennsylvania and possibly southeastern New York in Late Pennsylvanian time (Levine, 1986; Beaumont et al. 1987), the Findlay - Algonquin Arch was flexurally uplifted, following the tectonic model of Quinlan and Beaumont (1984; see fig. 5). This was similar to the proposed uplift of the Cincinnati Arch in response to thrusting in the Ouachitas and the southern Appalachians in the Mississippian - Pennsylvanian interval when clastic sediments eroded from the arch entered the western side of the Appalachian basin in Ohio and Pennsylvania, and subsequently as deltas into Illinois and Indiana by Middle Pennsylvanian time (Donaldson and Schumaker, 1981). Alleghanian thrusting migrated north to the central and northern Appalachians (Rodgers, 1967; Miller and Kent, 1988; Thomas and Schenk, 1988) possibly culminating in the Late Pennsylvanian tectonic loading of eastern Pennsylvania (Levine, 1986; Beaumont et al. 1987). We infer that Late Pennsylvanian basin-margin tectonic loading caused upwarp of the Findlay-Algonquin Arch which was eroded, supplying abundant sediments which filled the intervening flexural trough over west-central New York and areas to the west and northwest.

Other indirect evidence of a western provenance comes from the fault-bounded basins of the Canadian Maritime province. Paleocurrent data suggests that, for the Late Pennsylvanian - Early Permian siliciclastic sediments in these basins the most important source was far to the west (Thomas and Schenk, 1988). Could this source be connected to the western source - the Findlay - Algonquin Arch - proposed for western New York? We think this was certainly possible.

How long after the Devonian did sedimentation continue in the area? There is no easy answer to this question because the sedimentation and preservation rates of the missing strata cannot be correctly inferred. But, if the thrust blocks in eastern Pennsylvania were emplaced in the Late Pennsylvanian (Levine, 1986) and more than 4km of strata in southeast New York was derived from their subsequent erosion, then sedimentation in this part of New York State may have continued through much of the Permian. That a Permian source terrain was present somewhere in the east is also indicated by preserved Permian clastic rocks (the Dunkard Formation) of southwestern Pennsylvania which show evidence of an eastern provenance (Miller and Duddy, 1989) and the inferred former presence of 2 - 4km of Permian strata in the western half of Pennsylvania (Zhang and Davis, 1993; see fig. 8). Possibly, the thrust blocks in eastern Pennsylvania (shown as "Alleghanian Mountains" in fig. 24) were the common source of Permian sediments in Pennsylvania and New York.

How long did it take for 6km of post-Devonian strata to be deposited over west-central New York? It has been suggested that much sediment filling the trough in this area came from the uplifted Findlay - Algonquin Arch during the Late Pennsylvanian (fig. 24). Before the arch was uplifted and began shedding detrital sediments, the sedimentation rate over this area, particularly in Mississippian and Early Pennsylvanian times, must have been rather low because there was no important source of clastic sediments nearby. A lag time of several million years between thrust-load emplacement at the basin margin and uplift of the Findlay - Alogonquin Arch within the context of Quinlan and Beaumont's (1984) visco-elastic crustal model is possible. As the Findlay - Algonquin Arch rose, sediments were eroded from it probably at a much slower rate than the erosion of the the rapidly emplaced thrust blocks on the eastern side of the basin. Thus, it may have taken longer - perhaps up to late Permian - for the 6km-thick strata to be deposited over west-central New York.

The speculative paleogeographic and tectonic model for the study area during Late Pennsylvanian through the Permian (fig. 24) shows an inland basin with two elongate troughs separated by an area of thinner sediment accumulation. The basin is bounded in the north and northwest by highlands of probably moderate elevation along the Findlay - Algonquin Arch. The arch was uplifted due to peripheral upwarping in response to west- or northwest-verging overthrusting at the basin margin which probably overrode northeastern Pennsylvania (Levine, 1986) as well as parts of southeastern New York (Beaumont et al. 1987). These thrust blocks, here called "Alleghanian Mountains", bounded the basin to the southeast.

It is unlikely that during the Pennsylvanian - Permian interval this basin was covered by sea. First, the preserved Carboniferous rocks in Pennsylvania are fluvial in origin; no marine facies are known. Secondly, during the same time interval the region south of New York witnessed Alleghanian folding, overthrusting and uplift - a

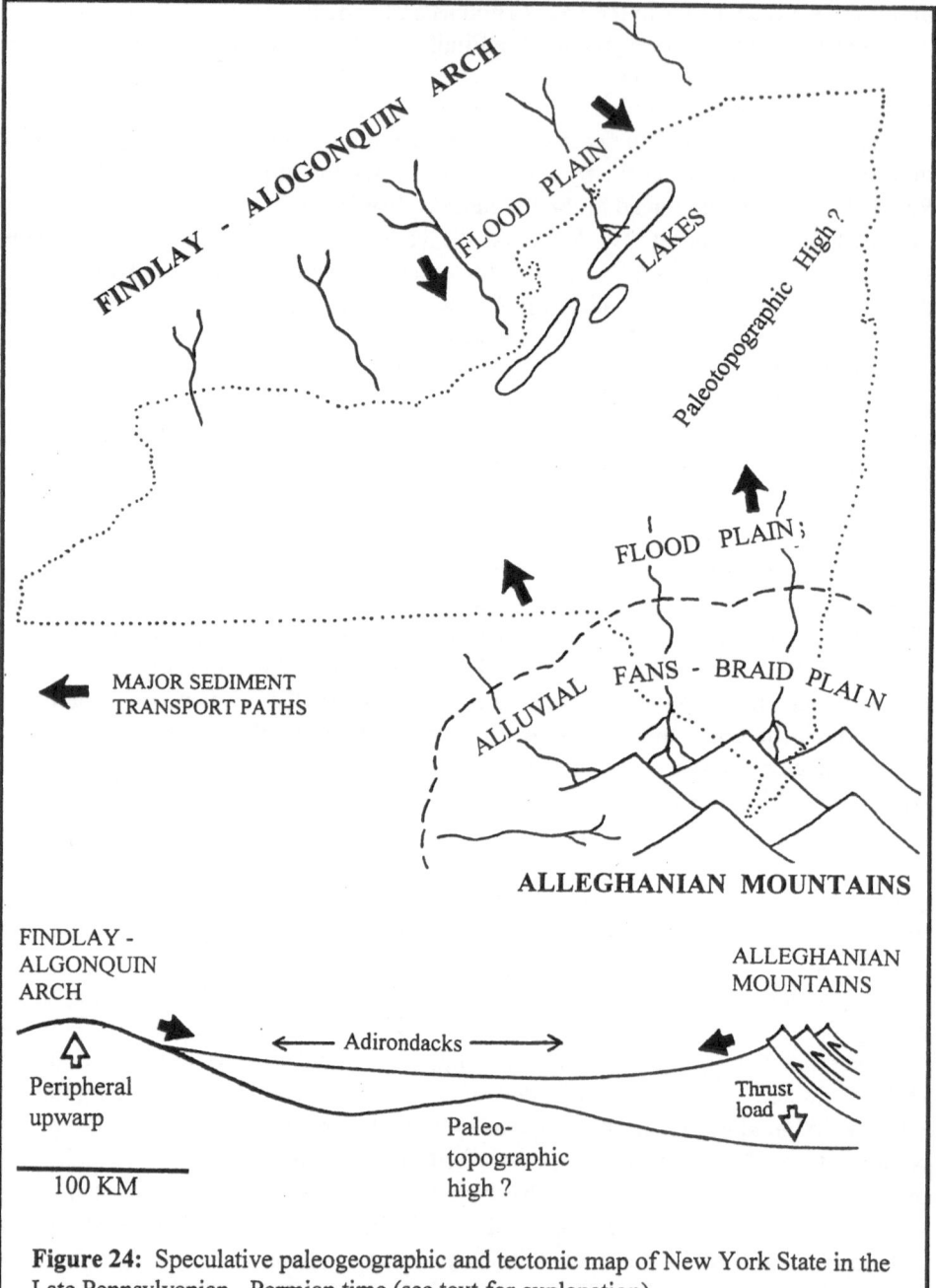

Figure 24: Speculative paleogeographic and tectonic map of New York State in the Late Pennsylvanian - Permian time (see text for explanation).

very unfavorable time for the sea to return to New York. Most probably, the inland basin was dotted with lakes or isolated remnants of the older Appalachian seaway. These may be compared with the ephemeral lakes in the foreland basin north of the Tien Shan range, China (Allen et al. 1991)

The Alleghanian Mountains (fig. 24), being mainly thrust generated, rose rapidly and were subjected to vigorous erosion. Clastic debris shed from these mountains was carried north, northwest and west by young rivers and was deposited as alluvial-fans, braided-river and flood-plain deposits which filled the trough over southeastern New York. If the Late Pennsylvanian - Permian climate in the region had become relatively dry, as Phillips and Peppers (1984) suggest, the consequent decrease in vegetation combined with high slope gradients were probably especially favorable for development of alluvial fans and braided river systems. The trough over southeastern New York was probably a visco-elastic response of the crust to loading by both the thrust sheets and sediment wedge (Quinlan and Beaumont, 1984; Tankard, 1986; Beaumont et al. 1987).

The trough over the west-central New York possibly partly inherited its geometry from the ancestral Appalachian basin, and was covered by lakes into which rivers of moderate gradients deposited sediments transported from the Findlay- Algonquin Arch. Besides river channel and flood plain deposits, small deltas might have formed, and in the eastern portion of this trough even evaporites might have been deposited. The cause of thinner post-Devonian sediment accumulation in a NE-elongate zone separating the two troughs is not clear. Probably, it marked the position of a paleotopographic high (fig. 24) that existed in post-Devonian time. This zone follows the axis of the Adirondack basement high along which the highest peaks of the present-day Adirondack Mountains are located. Over much of this zone, metanorthosite rocks are exposed (Isachsen and Fisher, 1970; Wiener et al. 1984, fig. 1). According to Whitney (1983), the relatively low-density metanorthosites were differentially uplifted over the surrounding rocks as "rigid domes" during and following the Grenville orogeny. Although, this took place in the Precambrian, it is possible that this portion of the Adirondacks survived as a topographic high of low amplitude or as a locus of renewed, local uplift(s) during the subsequent geologic time. There is stratigraphic evidence that the southern part of this zone (around location 63 in fig. 9 and 21) remained emergent in Middle Ordovician time during the deposition of carbonate sediments of the Black River and lower Trenton groups as indicated by depositional pinch-out of these rocks toward location 63 from both east and west (Fisher, 1977, plates 3, 4; pers. comm. 1990). It is possible that this zone acted as a topographic barrier again in the post-Devonian Period. The occurrence of the highest elevations of the present-day Adirondacks along this zone also suggests that the same low-density metanorthosites were probably again uplifted more than the surrounding rocks in the latest cycle of uplift.

If our interpretation of Late Pennsylvanian to Late Permian sediment accumulation of substantial thickness over New York State is correct, it follows that the uplift and erosion of the thick Paleozoic cover could have begun only toward the end of the

Permian. It also means that while Alleghanian movement folded and uplifted much of the central and southern Appalachians to the south, much of New York State remained, or transformed into, an inland basin and received sediments from the various components of the Alleghanian uplift: the uplift of New York probably lagged behind the Appalachian basin to the south significantly.

In our opinion, the subsequent uplift was probably initiated by pre-rifting doming of the eastern North America as it migrated over hot-spots or mantle plumes in the Triassic period (Burke et al. 1973; Morgan, 1981; Manspeizer, 1985). As doming of this nature generally involves large areas - as much as 700km across (Hay et al. 1981) - New York was probably riding a broad regional uplift, not a narrow hotspot track. However, it is difficult to place constraints on the subsequent uplift history of New York from available studies.

Apatite fission-track dates compiled by Crough (1981, fig. 1) show that areas adjacent to northeastern New York were uplifted through the "closure temperature" (100 - 110° C) of apatite between 115 and 120 Ma before present. This apparently happened for the basal Middle Devonian Gilboa sediments of the Catskill Mountains in eastern New York 124±5 Ma ago (Lakatos and Miller, 1983). Miller and Lakatos (1983) inferred from their fission-track study that the anorthosites of eastern Adirondacks were uplifted "at a slower rate between 147 and 86 Ma and at a faster rate since". Johnsson (1986) interpreted that the Middle Devonian Tioga bentonite was uplifted through the "closure temperature" of apatite 236±17 Ma ago in westernmost New York, 183±13 Ma ago in central New York, and 193±15 Ma ago in east-central New York. Miller and Duddy (1989) concluded that a regional uplift of New York State took place between 120 - 140 Ma before present. According to Isachsen (1985) the Adirondack Massif, probably a manifestation of a continental hot-spot epeirogeny - although the area does not show any abnormal heat flow -, began rising in the Tertiary and is currently experiencing an extraordinary uplift rate of 3.7 mm/year at the center of the dome.

These various studies give the impression that uplift of some part or other of New York has occurred since the Early Triassic and it is difficult to isolate periods of episodic uplift, if any. We are unable to provide any new insight into this problem from our data: clearly, more work is needed.

10 SUMMARY AND CONCLUSIONS

Paleotemperature assessment of rocks from sixteen rock units of New York State using fluid inclusions, organic maturation, clay diagenesis, and stable isotopes reveals that the surface and near-surface rocks experienced high paleo- temperatures often exceeding 200° C. The temperatures calculated from illite crystallinity index are believed to be closest to the actual maximum paleotemperatures experienced by the studied samples; whereas pressure-uncorrected maximum fluid-homogenization temperatures represent a minimum limit of the actual maximum paleotemperatures.

Temperatures calculated from maximum vitrinite reflectance (mean of upper 10% of the readings) correlate better with the illite crystallinity temperatures than mean vitrinite reflectance. Thermal alteration index and O-isotope derived temperatures were found to be the least usable for paleotemperature assessment.

The high paleotemperature signatures were attained during deep burial from normal regional geothermal heat in the northern Appalachian foreland basin. Burial depths calculated from both fluid-homogenization temperatures and illite- crystallinity temperatures indicate the former presence of significant thicknesses of post-Devonian strata across the New York State. Isopach maps constructed from the thicknesses of post-Devonian strata show the former presence of as much as 6 km of post-Devonian strata in southeastern and west-central New York separated by an area of thinner thickness.

Various geological considerations suggest that sedimentation in New York probably continued up to the Late Permian. It is speculated that the major source of sediments were the rapidly emplaced Alleghanian thrust sheets (Late Pennsylvanian - Early Permian) in the vicinity of the southeastern New York and the resulting flexural uplifted areas along the Findlay-Algonquin Arch, northwest of New York. Thus, the Late Paleozoic Alleghanian movement, unlike that in the southern and central Appalachians, appears to have marked a period of sedimentation in the New York region in which various Alleghanian uplifts provided the sediment sources. Eroded detritus from these sources was deposited in the adjacent enclosed foreland basin. Environments of deposition may have included alluvial-fans, braid-plains, flood-plains, and small deltas feeding ephemeral lakes.

Uplift of the area probably began toward the end of the Permian and was initiated by pre-rifting doming of eastern North America as it migrated over hot-spots or mantle plumes in the Triassic period. Various published studies suggest that uplift has continued in the region since the Triassic.

REFERENCES

Abercrombie, H.J., Hutcheon, I.E., Bloch, J.D., and Caritata, P.D., 1994, Silica activity and the smectite-illite reaction: Geology, v. 22, p. 539 - 542.

Allen, M.B., Windley, B.F., Chi, Z., Zhong-Yan, Z., and Guang- Rei, W., 1991, Basin evolution within and adjacent to the Tien Shan Range, NW China: Jour. Geol. Soc. Lond., v. 148, p. 369 - 378.

Anderson, T.F. and Arthur, M.A., 1983, Stable isotopes of oxygen and carbon and their application to sedimentologic and paleoenvironmental problems, p.1-11 -- 1-151, in Arthur, M.A., Anderson, T. F., Kaplan, I. R. et al. (eds.) Stable Isotopes in Sedimentary Geology: Soc. Econ. Paleon. Mineral. Short Course no. 10, 436p.

Ando, C.J., Cook, F.A., Oliver, J.E. et al., 1983, Crustal geometry of the Appalachian orogen from seismic reflection studies, p. 83 - 102, in Hatcher, R.D. Jr., Williams, H., and Zietz, I. (eds.) Contribution to the Tectonics and Geophysics of Mountain Chains: Geol. Soc. Amer. Memoir 158, 123p.

Arthur, M.A., Anderson, T.F., Kaplan, I.R., Veizer, J. and Land, L.S., 1983, Stable Isotopes in Sedimentary Geology: Soc. Econ. Paleon. Mineral. Short Course no. 13, 436p.

Barker, Ch.E., 1983, Influence of time on metamorphism of sedimentary organic matter in liquid-dominated geothermal systems, western North America: Geology, v. 11, p. 384 - 388.

_____, 1988, Temperature and time in thermal maturation of sedimentary organic matter, in Naeser, N.D. and McCulloh, T.H. (eds.), Thermal History of Sedimentary Basins, Springer-Verlag, New York, p. 73 -98.

_____, 1991, Implications for organic maturation studies of evidence for a geologically rapid increase and stabilization of vitrinite reflectance at peak temperature: Cerro Prieto geothermal system, Mexico: Amer. Assoc. Petrol. Geol. Bull., v. 75, p. 1852 - 1863.

_____ and Goldstein, R.H., 1990, Fluid-inclusion technique for determining maximum temperature in calcite and its comparison to the vitrinite reflectance geothermometer: Geology, v. 18, p. 1103 - 1106.

_____ and Pawlewicz, M.J., 1986, The correlation of vitrinite reflectance with maximum temperature in organic matter, in Buntebarth, G. and Stegena, L. (eds.) Paleogeothermics, lecture notes in earth sciences, v. 5, Springer-Verlag, Berlin, p. 79 - 93.

Barnaby, R.J. and Read, J.F., 1992, Dolomitization of a carbonate platform during late burial: Lower to Middle Cambrian Shaly Dolomite, Virginia Appalachians: Jour. Sed. Petrol., 1992, v. 62, p. 1023 - 1043.

Basu, R.A., Rubury, E., Mehnert, H., and Tatsumoto, M., 1984, Sm-Nd, K-Ar, and petrologic study of some kimberlites from the eastern United States and their implications for mantle evolution: Contrib. Mineral. Petrology, v. 86, p. 35 - 44.

Beaumont, C., Quinlan, G.M. and Hamilton, J., 1987, The Alleghenian Orogeny and its relationship to the evolution of the Eastern Interior, North America: Can. Soc. Pet. Geol. Mem. 12, p. 425 - 445.

Bedard, J.H., 1985, The opening of the Atlantic, the Mesozoic New England province, and mechanism of continental breakup: Tectonophysics, v. 113, p. 209 - 232.

Bodnar, R.J. and Bethke, P.M., 1984, Systematics of stretching of fluid inclusions I: fluorite and sphalerite at 1 atmosphere confining pressure: Econ. Geol. v. 79, p. 141 - 161.

Boles, J.R. and Franks, S.G., 1979, Clay diagenesis in Wilcox Sandstones of southwest Texas: Jour. Sed. Petrol., v. 49, p. 55 - 70.

Bosworth, W., Rowley, D.B., Kidd, W.S.F., and Steinhardt, C., 1988, Geometry and style of post-obduction thrusting in a Paleozoic orogen: the tectonic frontal thrust system: Jour. Geology, v. 96, p. 163 - 180.

Boucot, A.J., Brett, C.E., Oliver Jr., W.A., and Blodgett, R.B., 1986, Devonian faunas of the Sainte-Helene Island breccia, Montreal, Quebec, Canada: Can. Jour. Earth. Sci., v. 23, p. 2047 - 2056.

Brockerhoff, F.G. and Friedman, G.M., 1987, Paleo-depths of burial of Middle Ordovician carbonates in New York state and Vermont: Northeastern Geol., v. 9, p. 51 - 58.

Broughton, J.G., Fisher, D.W., Isachsen, Y.W., and Rickard, L.V., 1966, Geology of New York, a short account: N. Y. State Mus. And Sci. Serv. Educ. Leaflet 20, 45p.

Buchan, K.L. and Schwarz, E.J., 1987, Determination of maximum temperature profile across dyke contacts using remanent magnetization, and its applications, p.221-220, in Halls, H.C. and Fahrig, W.F. (eds.) Mafic Dyke Swarms, Geol. Assoc. Can. Spec. pap. 34, 503p.

Burke, K., Kidd, W.S., and Wilson, J.T., 1973; Relative and latitudinal motion of Atlantic hotspots: Nature, v. 245, p. 133 - 137.

Burruss, R.C., 1987, Diagenetic paleotemperatures from aqueous fluid inclusions: Re-equilibration of inclusions in carbonate cements by burial heating: Mineral. Magazine, v. 51, p. 477 - 481.

_____ , 1988, Paleotemperatures from fluid inclusions: advances in theory and technique, in Naeser, N.D. and McCulloh, T.H. (eds.), Thermal History of Sedimentary Basins, Methods and Case Histories, p. 119 - 131. Springer-Verlag, New York.

Bustin, R.M., 1989, Diagenesis in kerogen, *in* Hutcheon, I.E. (ed.) Short course in burial diagenesis: Mineral. Assoc. Can. Short Course Handbook, v. 15, p. 1 - 38.

Cardott, B.J. and Lambert, M.W., 1985, Thermal maturation by vitrinite reflectance of Woodford Shale , Anadarko Basin, Oklahoma: Amer. Assoc. Petrol. Geol. Bull., v. 69, p. 1982 - 1998.

Castano. J.D. and Sparks, D.M., 1974, Interpretation of vitrinite reflectance measurements in sedimentary rocks and determination of burial history using vitrinite reflectance and authigenic minerals: Geol. Soc. Amer. Spec. Pap. 153, p. 31 - 52.

Cathles, L.M. and Smith, A.T., 1983, Thermal constraints on the formation of Mississippi Valley-type lead-zinc deposits and their implications for episodic basin dewatering and deposit genesis: Econ. Geol., v. 78, p. 983 - 1002.

Choquette, P.W. and James, N.P., 1987, Diagenesis #12, Diagenesis in limestones - 3. The deep burial environment: Geosci. Canada, v. 14, p. 3 - 35.

Clark, T.H., 1972, Region de Montreal: Ministere des Richesses naturelles du Quebec, Service de l'exploration geologique, Rapport geologique - 154, 244p.

Clayton, G., 1989, Vitrinite reflectance data from the Kinsale Harbour - Old Head of Kinsale area, southern Ireland, and its bearing on the interpretation of the Munster Basin: Jour. Geol. Soc. Lond., v. 146, p. 611 - 616.

Clynne, M.A. and Potter, R.W. II, 1977, Freezing point depression of synthetic brines: Geol. Soc. Amer. Abst. with Programs, 9, p. 930.

Colten-Bradley, V.A., 1987, Role of pressure in smectite dehydration-effects on geopressure and smectite-to-illite transformation: Amer. Assoc. Petrol. Geol. Bull., v. 72, p. 1414 - 1427.

Connolly, C.A., 1989, Thermal history and diagenesis of the Wilrich Member shale, Spirit River Formation, northwest Alberta: Bull. Can. Petrol. Geol., v. 37, p. 182 - 197.

Cook, F.A., Albough, D.S., Brown, L.D., et al., 1979, Thin-skinned tectonics in the crystalline southern Appalachians: COCORP seismic reflection profiling of the Blue Ridge Piedmont: Geology, v. 7, p. 563 - 567.

COSUNA (Correlation of Stratigraphic Units of North America) Chart Series, 1985, Appalachian Region Set, Catalog no. 692, Amer. Assoc. Petrol. Geol.

Crelling, J.C. and Dutcher, R.R., 1980, Principles and applications of coal petrology, Great Lake Sect. Soc. Econ. Paleon. Mineral. Short Course, no. 8, 127p.

Crough, S.T., 1981, Mesozoic hotspot epeirogeny in eastern North America: Geology, v. 9, p. 2 - 6.

Daniels, E.J., Altaner, S.P., Marshak, S., and Eggelston, J.R., 1990, Hydrothermal alteration in anthracite from eastern Pennsylvania: Geology, v. 18, p. 247 - 150.

De Boer, J., McHone, J.G., Puffer, J.H., Ragland, P.C. and Whittington, D., 1988, Mesozoic and Cenozoic magmatism, *in* Sheriden, R.E. and Grow, J.A. (eds.) The Geology of North America, Vol. 1-2, The Atlantic Continental Margin, U.S., Geol. Soc. Amer. p. 217 - 241

Delaney, P.T., 1987, Heat transfer during emplacement and cooling of mafic dykes, p.31-46, *in* Halls, H.C. and Fahrig, W.F. (eds.) Mafic Dyke Swarms, Geol. Assoc. Can. Spec. Pap. 34, 503p.

Deming, D., 1992, Catastrophic release of heat and fluid in continental crust: Geology, v. 20, p. 83 - 86.

_____ and Nunn, 1991, Numerical simulation of brine migration by topographically driven recharge: Jour. Geophys. Res., v. 96, p. 2485 - 2499.

_____, Sass, J.H., Lachenbruch, A.H., and De Rito, R.F., 1992, Heat flow and subsurface temperature as evidence for basin-scale ground-water flow, North Slope of Alaska: Geol. Soc. Amer. Bull., v. 104, p. 528 - 542.

Dennison, J.M., 1983, Comment on "Tectonic model for kimberlite emplacement in the Appalachian plateau of Pennsylvania: Geology, v. 11, p. 252 - 253.

De Vivo, B. And Frezzotti, M.L. (eds.), 1994, Fluid Inclusion in Minerals: Methods and Applications: Virg. Polytech. Inst. And Univ., Blacksburg, VA, USA, 376p.

Donaldson, A.C. and Schumaker, R.C., 1981, Late Paleozoic molasse of central Appalachian, *in* Miall, A.D. (ed.), Sedimentation and Tectonics in Alluvial Basins: Geol. Assoc. Can. Spec. Pap. 23, p. 99 - 124.

Dow, W.G. and O'Connor, D.I., 1982, Kerogen maturity and type by reflected light microscopy applied to petroleum exploration: How to Asses Maturation and Paleotemperatures, Soc. Econ. Paleon. Mineralogists Short Course, No. 7, p. 133-158.

Duba, D. and William-Jones, A.E., 1983, The application of illite crystallinity, organic matter reflectance and isotopic techniques to mineral exploration: a case study in southwestern Gaspe, Quebec: Econ. Geol., v. 78, p. 1350 - 1363.

Dunning, G.R. and Hodych, J.D., 1990, U-Pb zircon and baddeleyte age for the Palisade and Gettysburg sills of northeastern United States: implications for the age of Triassic - Jurassic boundary: Geology, v. 18, p. 151 - 164.

Duddy, I.R., Gleadow, A.J.W., Green, P.F. et al., 1987, Quantitative estimates of thermal history and maturation using AFTA (Apatite Fission Track Analysis) in extensional and foreland basins - selected case studies: Amer. Assoc. Petrol. Geol. Bull., v. 71, p. 550 - 551.

Eberl, D., 1978, The reaction of montmorillonite to mixed-layer clay: the effect of interlayer alkali and alkaline earth cations: Geochim. Cosmochim. Acta, v. 42, p. 1 - 7.

Edmunds, W.E., Berg, T.M., Sevon, W.D., Piotrowski, R.C., Heyman, L. and Rickard, L.V., 1979, The Mississippian and Pennsylvanian systems in the United States - Pennsylvania and New York: U.S. Geol. Surv. Prof. Paper 1110-B, 33p.

Epstein, A.G., Epstein, J.B. and Harris, L.D., 1977, Conodont color alteration - an index to organic metamorphism: US Geol. Surv. Prof. Pap. 995, 27p.

Epstein, J.B. and Lyttle, 1987, Structure and stratigraphy above, below and within the Taconic unconformity, southeastern New York: Field trip guidebook, NY State Geol. Assoc., 59th Ann. Meetg., p. C1 - C7.

Engelder, T. and Engelder, R., 1977, Fossil distortion and decollement tectonics on the Appalachian plateau: Geology, v. 5, p. 457 - 476.

_____ and Geiser, P.A., 1979, The relationship between pencil cleavage and lateral shortening within the Devonian section of the Appalachian plateau, New York: Geology, v. 5, p. 460 - 464.

_____ and _____, 1980, On the use of regional joint sets as trajectories of paleostress fields during the development of the Appalachian plateau, New York: Jour. Geophys. Res., v. 85, p. 6319 - 6341.

Eslinger, E. and Pevear, D., 1988, Clay Mineralogy for Petroleum Geologists and Engineers: Soc. Econ. Paleon. Mineralogists Short Course Notes, no. 22, 415p.

Ettensohn, F.R., 1985a, The Catskill delta complex and the Acadian orogeny: a model, p. 39 - 50, in Woodrow, D.L. and Sevon, W.D. (eds.) The Catskill Delta: Geol. Soc. Amer. Spec. Pap. 201, 246p.

_____, 1985b, Controls on development of Catskill delta complex basin facies, p. 65 - 78, in Woodrow, D.L. and Sevon, W.D. (eds.) The Catskill Delta: Geol. Soc. Amer. Spec. Pap. 201, 246p.

Faill, R.T., 1985, The Acadian orogeny and the Catskill delta, p. 15 - 38, in Woodrow, D.L. and Sevon, W.D. (eds.) The Catskill Delta: Geol. Soc. Amer. Spec. Pap. 201, 246p.

Fakundiny, R.H., Cadwell, D.H., and Fleisher, P.J., 1989, Geology of Wine County of New York: Field trip guidebook, Internat. Geol. Congress 28th meetg., 64p.

Fisher, D.W., 1977, Correlation of the Hadrynian, Cambrian and Ordovician Rocks in New York State: New York State Museum Map and Chart Series No. 25.

Foster, B.P., 1970, A study of the kimberlite - alnoite dikes in central New York: Unpub.M.S. thesis, State Univ. of N. Y., Buffalo, 55p.

Frey, M., Teichmuller, M., Teichmuller, R., Mullins, J., Kunzi, B., Breitschmid, A., Gruner, U., and Schwizer, B., 1980, Very low-grade metamorphism in external parts of the Central Alps: illite crystallinity, coal rank and fluid-inclusion data: Eclogae Geol. Helv., v. 73, p. 173 - 203.

Friedman, G.M., 1959, Identification of carbonate minerals by staining methods: J. Sed. Petrol., v. 29, p. 87 - 97.

_____, 1965, Terminology of crystallization textures and fabrics in sedimentary rocks: Jour. Sed. Petrol., v. 35, p. 643 - 655.

_____, 1987a, Deep-burial diagenesis: its implication for vertical movements of the crust, uplift of the lithosphere and isostatic unroofing - a review: Sed. Geology, v. 50, p. 67 - 94.

_____, 1987b, Vertical movement of the crust; case histories from the northern Appalachian Basin: Geology, v. 15, p. 1130 - 1133.

_____, 1988, The Catskill tectonic fan-delta complex: northern Appalachian basin: Northeastern Geology, v. 10, p. 254 - 257.

_____, 1990, Anthracite and concentrations of alkaline feldspar (microcline) in flat-lying undeformed Paleozoic strata: a key to large-scale vertical crustal uplift, p. 16 - 28, *in* Heling, D. Rothe, P., Forstner, U., and Stoffers, P. (eds.), Sediments and sedimentary environments: Berlin, Springer - Verlag, 317p.

_____, 1994, Pangean orogenic and epeirogenic uplifts and their possible climatic significance: Geol. Soc. Amer. Spec. Pap. 288, p. 159 - 167.

_____ and Sanders, J.E., 1982, Time-temperature-burial significance of Devonian anthracite implies former great (6.5km) depth of burial of Catskills Mountains, New York: Geology, v. 10, p. 93 - 96.

_____ and _____, 1983, Reply on "Time - temperature-burial significance of Devonian anthracite implies former great (~6.5km) depth of burial of Catskill Mountains, New York": Geology, v. 11, p. 123 - 124.

_____, _____, and Martini, P., 1982, Excursion 17A: Sedimentary facies: products of sedimentary environments in a cross-section of the classic Appalachian mountains and adjoining Appalachian basin in New York and Ontario: Internat. Assoc. Sediment. 11th Internat. Conf. field excursion guidebook, 274p.

Friedman, I. and O'Neil, J.R., 1977, Compilation of stable isotope fractionation factors of geochemical interest, in Fleischer, M. (ed.) Data on Geochemistry, 6th ed., U.S. Geol. Surv. Prof. Pap., 440 KK.

Fritz, P. and Smith, D.G.W., 1970, The isotopic composition of secondary dolomite: Geochim. Cosmochim. Acta, v. 34, p. 1161 - 1173.

Gale, P.E. and Siever, R., 1986, Diagenesis of Middle to Upper Devonian Catskill facies sandstones in southeastern New York: Amer. Assoc. Petrol. Geol. Bull., v. 70, p. 592 - 593 (abst.).

Garven, G., 1989, A hydrological model for the formation of giant oil sands deposits of Western Canadian Sedimentary Basin: Econ. Geology, v. 289, p. 105 - 166.

_____ and Freeze, A.R., 1984, Theoretical analysis of role of groundwater flow in the genesis of stratabound ore deposits I. Mathematical and numerical model. 2. Quantitative results: Amer. Jour. Sci., v. 284, p. 1085 - 1174.

Geiser, P. And Engelder, T., 1983, The distribution of layer parallel shortening fabrics in the Appalachian foreland of New York and Pennsylvania: evidence of two non-coaxial phases of the Alleghanian orogeny, p. 161 - 177, *in* Hatcher, R.D. Jr., Harold, W., and Zietz, I. (eds.) Contribution to the tectonics and geophysics of mountain chains: Geol. Soc. Amer. Bull., v. 101, p. 221 - 230.

Gerlach, J.B., 1987, Post-Devonian burial history of the New York state Appalachian Basin based on Lopatin modeling of regional vitrinite reflectance trends: Unpub. M.S. thesis, State Univ. of NY, Stony Brook, 135p.

Gillette, T., 1947, The Clinton of western and central New York: New York Museum Bull. 341, 191p.

Goldstein, R.H., 1986, Reequilibrium of fluid inclusions in low-temperature calcium carbonate cement: Geology, v. 14, p. 792 - 795.

_____, 1988, Cement stratigraphy of Pennsylvanian Holder Formation, Sacramento Mountains, New Mexico: Amer. Assoc. Petrol. Geol., v. 72, p. 425-438.

_____ and Reynolds, T.J., 1994, Systematics of fluid inclusions in diagenetic minerals: SEPM (Soc. Sedim. Geol.) Short course no. 31, 212p.

Gregg, J.M. and Sibley, D.F., 1984, Epigenetic dolomitization and the origin of xenotopic dolomite texture: Jour. Sed. Petrol., v. 54, p. 908 - 931.

Gross, M.R., Engelder, T. And Poulson, S.R., 1992, Veins in the Lockport dolostone: evidence for an Acadian fluid circulation system: Geology, v. 20, p. 971 - 974.

Gurney, G.G. and Friedman, G.M., 1987, Burial history of the Devonian Cherry Valley carbonate sequence, Cherry valley, New York: Northeastern Geol., v. 9, p. 1 - 11.

Guthrie, J.M., Houseknecht, D.W. and Johns, W.D., 1986, Relationships among vitrinite reflectance, illite crystallinity, and organic geochemistry of Carboniferous strata, Ouachita Mountains, Oklahoma and Arkansas: Amer. Assoc. Petrol. Geol. Bull., v. 70, p. 26 - 33.

Gwinn, V.E., 1964, Thin-skinned tectonics in the Plateau and northwestern Valley and Ridge provinces of the central Appalachians: Geol. Soc. Amer. Bull., v. 75, p. 863 - 900.

Haas, J.L.Jr., 1976, Physical properties of the co-existing phases and the thermo-chemical properties of the H_2O component in boiling NaCl solutions: U.S. Geol. Surv. Bull. 1421-A, 73p.

Hanor, J.S., 1987, Origin and Migration of Subsurface Sedimentary Brines: Soc. Econ. Paleon. Mineral. Short Course no. 21, 247p.

Harris, A.,1979, Conodont color alteration: an organo- metamorphic index and its application to Appalachian Basin Geology, *in* Scholle, P.A. and Schluger, P.R. (eds.) Aspects of Diagenesis: Soc. Econ. Paleon. Mineral. Spec. Pub., 26, p. 3 - 16.

_____, A., Harris, L.D. and Epstein, J.B., 1978, Oil and gas data from Paleozoic rocks in the Appalachian basin: maps for assessing hydrocarbon potential and thermal maturity (conodont color alteration isograds and overburden isopachs): U.S. Geol. Surv. Misc. Investig. Map. I-917-E.

Harrison, W.E, Luza, K.V., Prater, M.L. and Cheung, P.K., 1983, Geothermal resource assessment in Oklahoma: Oklahoma Geol. Surv. Spec. pub., 83-1, 44p.

Hatcher, R.D. Jr., 1978, Tectonics of the western Piedmont and Blue Ridge , southern Apppalachians: review and speculation: Amer. Jour. Sci., v. 278, p. 276 - 304.

_____ and Zietz, I., 1980, Tectonic implications of regional aeromagnetic and gravity data from the southern Appalachians, in Wones, D.R. (ed.) The Caledonides of the U.S.A.: Blacksburg, Virg. Polytech. Inst. And State Uni. Mem. 2, p. 235 - 244.

Hay, W.W., Barron, E.J., Sloan, J.L., and Southam, J.R., 1981, Continental drift and the global pattern of sedimentation: Geol. Rundschau, v. 70, p. 302 - 315.

Heckel, P.H., 1973, Nature, origin and significance of Tully Limestone, an anomalous unit in the Catskill Delta, Devonian of New York: Geol. Soc. Amer. Spec. Pap. 138, 244p.

Heroux, Y., Chagnon, A. and Bertrand, R., 1979, Compilation and correlation of major thermal maturation indicators: Amer. Assoc. Petrol. Geol. Bull., v. 63, p. 2128 - 2144.

Hiscott, R.N., Pickering, K.T., and Beeden, D.R., 1986, Progressive filling of a confined Middle Ordovician foreland basin associated with Taconic orogeny, Quebec, Canada, in Allen, P.A. and Homewood, P. (eds.) Foreland Basins: Internat. Assoc. Sedimentologists Spec. Pub. No. 8, p. 309 - 325.

Hitchon, B., 1984, Geothermal gradients, hydrodynamics, and hydrocarbon occurrences, Alberta, Canada: Amer. Assoc. Petrol. Geol. Bull., v. 68, p. 713 - 743.

_____ and Friedman, I, 1969, Geochemistry and origin of formation waters in the western Canada sedimentary basin I: Stable isotopes of hydrogen and oxygen: Geochim. Cosmochim. Acta, v. 33, p. 1321 - 1349.

Hodge, D.S., 1984, Heat flow and subsurface temperature distributions in central and western New York: N.Y. State Energy Research and Development Authority Rep. 84-8.

_____, Eckert, R., and Raveta, F., 1982, Geophysical signatures of central and western New York State: Field trip guidebook, NY State Geol. Assoc. 54th ann. meetg., p. 3 - 18.

Hoffman, J. and Hower, J., 1979, Clay mineral assemblages as low grade metamorphic geothermometers - application to the thrust-faulted disturbed belt of Montana,USA, in Scholle, P.A. and Schluger, P.R. (eds.), Aspects of Diagenesis: Soc. Econ. Paleon. Mineral. Spec. Pub. no. 26, p. 55 - 79.

Hollister, L.S., 1981, Techniques for analyzing fluid inclusions in Hollister, L.S. and Crawford, M.L. (eds.) Short Course in Fluid Inclusions: application to petrology: Toronto, Mineral. Assoc. Can., p. 272 - 277.

_____, L.S. and Crawford, M.L. (eds.), 1981, Short Course in Fluid Inclusions: application to petrology: Toronto, Mineral. Assoc. Can., 304p.

Horvath, F., Dovenyi, P., Szalay, A., and Royden, L.H., 1988, Subsidence, thermal and maturation history of the Great Hungarian Plain, in Royden, L.H. and Horvath, F., (eds.) The Pannonian basin - a study in basin evolution: Amer. Assoc. Petrol. Geol. Bull., v. 45, p. 355 - 372.

Hood, A., Gutjahr, C.C.M. and Heacock, R.L., 1975, Organic metamorphism and the generation of petroleum: Amer. Assoc. Petrol. Geol. Bull., v. 59, p. 986 - 996.

Houseknecht, D.W. and Matthews, S.M., 1985, Thermal maturity of Carboniferous strata, Ouachita Mountains: Amer. Assoc. Petrol. Geol. Bull., v. 69, p. 335 - 345.

Hower, J, 1981, Shale diagenesis, clays and the resource geologist, in Longstaffe, F.J., (ed.), Short Course Handbook: Mineral. Assoc. Can., p. 199.

Hower, J., Eslinger, E., Hower, M.E., Perry, E.A., 1976, Mechanism of burial metamorphism of argillaceous sediments: I. Mineralogical and chemical evidence: Geol. Soc. Amer. Bull., v. 87, p. 725 - 737.

Hower, J.C. and Davis, A., 1981, Application of vitrinite reflectance anisotropy in the evaluation of coal metamorphism: Geol. Soc. Amer. Bull., v. 92, p. 350 - 366.

Hurley, N.F. and Lohman, K.C., 1989, Diagenesis of Devonian reefal carbonates in the Oscar range, Canning Basin, Western Australia: J. Sed. Petrol., v. 59, p. 127 - 146.

Inoue, A., Kohyama, N., Kitagawa, R., and Watanabe, T., 1987, Chemical and morphological evidence for the conversion of smectite to illite: Clays and Clay Minerals., v. 35, p. 111 - 120.

Isachsen, Y.W., 1985, Structural and tectonic studies in New York State: N.Y. State Geol. Surv., prepared for Div. of Radiation Programs and Earth Sciences, Office of Nuclear Regulatory Research, Washington, 74p.

_____, 1992, The Adirondacks: still rising after all these years: Nat. History, v. 5, p. 383 - 387.

_____ and Fisher, D.W. 1970: Geologic Map of New York, Adirondack sheet, N .Y. State Mus. Sci. Serv. no. 15.

_____, Landing, E., Lauber, J.M. et al. (eds.), 1991, Geology of New York: a simplified account: NY State Mus. Educational Leaflet 28, 284p.

Issler, D.R., 1984, Calculation of organic maturation levels for offshore Canada- implications for general application of Lopatin's method: Can. J. Earth Sci.,v. 21, p. 477 - 488.

Jackson, M., McCabe, C., Ballard, M.M. and Van der Voo, R., 1988, Magnetite authigenesis and diagenetic paleotemperatures across the northern Appalachian basin: Geology, v. 16, p. 592 - 595.

Jackson, M.L., 1979, Soil Chemical Analysis -- Advanced Course, 2nd. ed., published by author, Madison, Wisconsin, 580p.

Jacobi, C.H. and Dellwig, L.F., 1974, Appalachian foreland thrusting in Salina salt, Watkins Glen, New York, in Coogen, A.H. (ed.) Fourth Symposium on Salt, Northern Ohio Geol. Soc., Cleavland, p. 227 - 251

Jacobi, R.D., 1981, Peripheral bulge - a causal mechanism for the Lower/Middle Ordovician unconformity along the western margin of the northern Appalachians: Earth and Planet. Sci. Letters, v. 56, p. 245 - 251.

Johnson, K.G. and Friedman, G.M., 1969, The Tully clastic correlatives (Upper Devonian) of New York: a model for recognition of alluvial, dune (?), tidal nearshore (bar and lagoon), and offshore environments in a tectonic delta complex: Jour. Sed. Petrol., v. 39, p. 451 - 485.

Johnsson, M., 1985, Late Paleozoic - middle Mesozoic uplift rate, cooling rate and geothermal gradient for south-central New York state: Nuclear Tracks, v. 10, p. 295 - 301.

_____, 1986, Distribution of maximum burial temperatures across the northern Appalachian Basin and implication for Carboniferous sedimentation patterns: Geology, v. 14, p. 383 - 387.

_____, Howell, D.G., and Bird, K. J., 1993, Thermal maturity patterns in Alaska: implications for tectonic evolution and hydrocarbon potential: Amer. Assoc. Petrol. Geol. Bull., v. 77, p. 1874 - 1903.

Jones, P.H. and Wallace Jr., R.H., 1974, Hydrogeologic aspects of structural deformation in the northern Gulf of Mexico Basin: J. Res., U.S. Geol. Surv., v.2, p. 511 - 517.

Karig, D.E., 1987, Comments on "Distribution of maximum burial temperatures across northern Appalachian Basin and its implications for Carboniferous sedimentation patterns": Geology, v. 15, p. 278 - 279.

Kaufman, J., Meyers, W.J., and Hanson, G.N., 1990, Burial cementation in the Swan Hills Formation (Devonian), Rosevear Field, Alberta, Canada: Jour. Sed. Petrol., v. 60, p. 918 - 939.

Kay, S.M. and Foster, B.P., 1986, Kimberlites of the Finger Lakes region: NY State Geol. Assoc. 58th Annl. Meetg. Field Trip Guidebook, p. 219 - 238.

_____, Snedden, W.T., Foster, B.P., and Kay, R.W., 1983, Upper mantle and crustal fragments in the Ithaca kimberlites: Jour. Geology, v. 91, p. 277 - 290.

Keith, B.D. and Friedman, G.M., 1977, A slope-fan-basin-plain model, Taconic Sequence, New York and Vermont: Jour. Sed. Petrol., v. 47, p. 1220 - 1241.

Kent, D.V., 1985, Paleocontinental setting for the Catskill Delta, in Woodrow, D.L. and Sevon, W.D. (eds.) The Catskill Delta: Geol. Soc. Amer. Spec. Pap. 201, p. 9 - 13.

Kisch, H.J., 1980, Incipient metamorphism of Cambro-Silurian clastic rocks from the Jamtland Supergroup, central Scandinavia Caledonides, western Sweden: Illite crystallinity and vitrinite reflectance: Jour. Geol. Soc. London, v. 137, p. 271 - 288.

Kramers, J.W. and Friedman, G.M., 1986, The Centerfield Limestone and its clastic correlatives: Northeastern Geology, v. 8, p. 66 - 90.

Kreidler, W.L., Van Tyne, A.M. and Jorgensen, K.M., 1972, Deep Wells in New York State: N.Y. State Museum and Sci. Serv. Bull. no. 418A, 335p.

Kubler, B., 1968, Evaluation quantitative du metamorphisme par la cristallinite de l'illite; etat des progres realises annees: Bull. Centre due Recherches de Pau, v. 2, p. 385 - 397.

Lacazette, A.J., 1991, Natural hydraulic fracturing in the Bald Eagle sandstone in central Pennsylvania and the Ithaca siltstone at Watkins Glen, New York: Ph.D. Thesis, Pennsyl. State Univ., 255p.

Lakatos, S. and Miller, D.S., 1983, Fission-track analysis of apatite and zircon defines a burial depth of 4 to 7km for lowermost Upper Devonian Catskill Mountains, New York: Geology, v. 11, p. 103 - 104.

Land, L.S., 1980, The isotopic and trace element geochemistry of dolomite: the state of the art, in Zenger, D.H., Dunham, J.B., and Ethington, R.L. (eds.) Concepts and Models of Dolomitization: Soc. Econ. Paleon. Mineral. Spec. Pub. 28, p. 87 - 110.

_____, 1983, The application of stable isotopes to studies of the origin of dolomite and to problems of diagenesis of clastic sediments, p.4-1--4-22, in Arthur, M.A., Anderson, T.F., Kaplan, I. et al. (eds.) Stable Isotopes in Sedimentary Geology: Soc. Econ. Paleon. Mineral. Short Course no.10, 436p.

Landing, E., 1991, A view from the Hudson, Hudson - Mohawk lowlands and Taconic mountains, p. 53 - 66, in Isachsen, Y.W., Landing, E. Lauber, L.V. (eds.)

Geology of New York, a simplified account: NY State Museum, State Educ. Dept., 284p.

Lavoie, D. and Bourque, P.-A., 1993, Marine, burial, and meteoric diagenesis of early Silurian carbonate ramps, Quebec Appalachians, Canada: Jour. Sed. Petrol., v. 63, p. 233 - 2 47.

Lee, I.Y. and Friedman, G.M., 1987, Deep-burial dolomitization in the Ellenburger Group carbonates, West Texas and southeastern New Mexico: Jour. Sed. Petrol., v. 57, p. 544 - 557.

Lee, M-K., and Bethke, C.M., 1994, Groundwater flow, late cementation, and petroleum accumulation in the Permian Lyons Sandstones, Denver Basin: Amer. Assoc. Petrol. Geol. Bull., v. 78, p. 217 - 237.

Legall, F.D., Barnes, C.R., and MacQueen, R.W., 1981, Thermal maturation, burial history and hotspot development, Paleozoic strata of southern Ontario - Quebec, from conodont and acritarch color alteration studies: Bull. Can. Petrol. Geol., v. 29, p. 492 - 539.

Levine, J.R., 1983, Comment on "The-temperature-burial significance of Devonian anthracite implies former great (~6.5 km) depth of burial of Catskill Mountains, New York": Geology, v. 11, p. 122 - 123.

Levine, J.R., 1986, Deep burial of coal bearing strata, Anthracite region, Pennsylvania: sedimentation or tectonics: Geology, v. 14, p. 577 - 580.

Lopatin, N.V., 1971, Temperature and geologic time as factors in coalification: Akademiya Nauk SSSR Izvestiya, Seriya Geologicheskaya, no. 3, p. 95 - 106.

Machel, H. and Mountjoy, E.W., 1986, Chemistry and environments of dolomitization - a reappraisal: Earth Sci. Review, v. 23, p. 175 - 222.

Manspeizer, W., 1985, Early Mesozoic history of the Atlantic passive margin, in Poag. C.W. (ed.) Geologic Evolution of the United States Atlantic Margin, Van Nostrand Reinhold Co. Inc., p. 1 - 23.

_____ and Cousminer, H.L., 1988, Late Triassic - Early Jurassic synrift basins of the U.S. Atlantic margin, in Sheriden, R.E. and Grow J.A. (eds.) The Geology of North America, vol. 1 - 2, The Atlantic Continental Margin, U.S.: Geol. Soc. Amer. P. 197 - 216.

Marshak, S., 1986, Structure and tectonics of the Hudson Valley fold-thrust belt, eastern New York State: Geol. Soc. Amer. Bull., v. 97, p. 354 - 368.

McClelland-Brown, E., 1981, Paleomagnetic estimates of temperatures reached in contact metamorphism: Geology, v. 9, p. 112 - 116.

McHone, J.G., 1987, Cretaceous intrusions and rift features in the Champlain Valley of Vermont: Field trip guidebook, New Eng. Intercollegiate Geol. Conf. 79th ann. meetg., p.237 - 253.

McKenzie, D., 1981, The variation of temperature with time and hydrocarbon maturation in sedimentary basins formed by extension: E. Planet. Sci. Letters, v. 55, p. 87 - 98.

Meckel, L.D., 1970, Paleozoic alluvial deposition in central Appalachians: a summary in Fisher et al. (eds.) Studies of Appalachian Geology, Central and Southern: New York Interscience, p. 49 - 67.

Miller, D.S., 1990, Geothermometry of NY Upper Devonian sediments based on fission-track analysis of apatite and zircon: Abst. with Programs, Geol. Soc. Amer. NE sect. ann. meetg., p. 57.

_____ and Duddy, I.R., 1986, Burial and uplift history of northern Appalachian basin: apatite and zircon fission-track chronology: Abst. with Programs,Geol. Soc. Amer. NE sect. ann. meetg., p. 55.

_____ and _____., 1989, Early Cretaceous uplift and erosion of the northern Appalachian Basin, New York, based on apatite fission track analysis: Earth Planet. Sci. Letters, v. 93, p. 35 - 49.

_____ and Lakatos, S., 1983, Uplift rate of Adirondack anorthosite measured by fission-track analysis of apatite: Geology, v. 11, p. 284 - 286.

Miller, J.D. and Kent, D.V., 1989, Paleomagnetism of selected Devonian age plutons from Maine, Vermont, and New York: Northeastern Geol., v. 11, p. 66 - 76.

Moore, C.H., 1985, Upper Jurassic subsurface cements: a case history, p. 291-308, *in* Schneidermann, H. and Harris, P.M. (eds.) Carbonate Cements: Soc. Econ. Paleon. Mineral. Spec. Pub. 36, 379p.

Moore, D.M. and Reynolds, R.C. Jr., 1989, X-ray Diffraction and the Identification and Analysis of Clay Minerals: Oxford Univ. Press, Oxford, New York, 332p.

Morgan, W.J., 1981, Hotspot tracks and the opening of the Atlantic and Indian oceans, *in* Emiliani, C. (ed.), The Sea, Vol. 7, The Oceanic Lithosphere, John Wiley, NY, p. 443 - 487.

Morrow, D. W. And Issler, D.R., 1993, Calculation of vitrinite reflectance from thermal histories: a comparison of some methods: Amer. Assoc. Petrol. Geol. Bull., v. 77, p. 610 - 624.

Mose, D.G., Ratcliffe, N.M., Odom, A.L., and Hayes, J., 1976, Rb-Sr geochronology and tectonic setting of the Peekskill pluton, southeastern New York: Geol. Soc. Amer. Bull., v. 87, p. 361 - 365.

Nadeau, P.H., Wilson, M.J., McHardy, W.J., and Tait, J.M., 1984, Interparticle diffraction: a new concept for interstratified clays: Clays and Clay Minerals, v. 19, p. 757 - 769.

Northrop, D.H. and Clayton, R.N., 1966, Oxygen isotope fractionation in systems containing dolomite: Jour. Geology, v. 74, p. 174 - 196

Okran, N. And Voight, B., 1985, Regional joint evolution in the Valley and Ridge province of Pennsylvania in relation to the Alleghanian orogeny *in* Gold, D.P. (ed.) Central Pennsylvania geology revised: Penn. Geol. Surv. Ann. field conf. of Penn. geol. guidebook, p. 144 - 163.

Oliver, J., 1986, Fluids expelled tectonically from orogenic belts: their role in hydrocarbon migration and other geologic phenomena: Geology, v. 14, p. 99-102.

Olsen, P.E., McCune, A.R., and Thompson, K.S., 1982, Correlation of Early Mesozoic Newark Supergroup (eastern North America) by vertebrates, especially fishes: Amer. Jour. Sci., v. 282, p. 1 - 44.

O'neil , J.R. and Epstein, S., 1966, Oxygen isotope fractionation in system dolomite -calcite -carbon dioxide: Science, v. 152, p. 198 - 201.

Osberg, P.H., 1988, Silurian to Lower Carboniferous tectonism in the Appalachians of the U.S.A., *in* Harris, A.L. and Fettes, D.J. (eds.) The Caledonian - Appalachian orogen: Geol. Soc. Spec. Pub. No. 38, p. 449 - 452.

Paxton, S.T., 1983, Relationships between pennsylvanian-age lithic sandstone and mudrock diagenesis and coal rank in the central Appalachian: Ph.D. thesis, Penn. State Univ., Univ. Park, Penn., 503p.

Pelletier, B.R., 1958, Pocono paleocurrents in Pennsylvania and Maryland: Geol. Soc. Amer. Bull., v. 69, p. 1033 - 1064.

Perry, E.A. and Hower, J., 1970, Burial diagenesis in Gulf Coast pelitic sediments: Clays and Clay Min., v. 18, p. 165 - 177.

Perry, W.J. Jr., 1978, Sequential deformation in the central Appalachians: Amer. Jour. sci., v. 278, p. 518 - 542.

Pevear, D.R., Williams, V.E., and Mustoe, G.E., 1980, Kaolinite, smectite, and K-rectorite in bentonites: relation to coal rank at Tulameen, British Columbia: Clays and Clay Minerals, v. 28, p. 241 - 254.

Phillips, T.L. and Peppers, R.A., 1984, Changing patterns of Pennsylvania coal -swamp vegetation and implication of climatic control on coal occurrence: Internat. Jour. Coal Geol., v. 3, p. 205 - 255.

Pollastro, R.M. and Barker, Ch. E., 1986, Application of clay- mineral, vitrinite reflectance, and fluid inclusion studies to the thermal burial history of the Pinedale Anticline, Green River Basin, Wyoming, *in* Gautier, D.L. (ed.) Roles of Organic Matter in Sediment Diagenesis, Soc. Econ. Paleon. Mineral. Spec. Pub. no. 38, p. 73 - 84.

Poole, W.H., Sanford, B.V., Williams, H., and Kelley, D.G., 1970, Chart II: Geotectonic correlation chart for southeastern Canada, *in* Douglas, R.J.W. (ed.) Geologic and economic minerals of Canada, maps and charts.

Potter, R.W.II, 1977, Pressure corrections for fluid inclusion homogenization temperatures based on volumetric properties of system NaCl-H$_2$O: Jour. Res., U.S. Geol. Surv., 5, p. 603 - 607.

_____ and Brown, D.L., 1975, The volumetric properties of aqueous sodium chloride solutions from 0 to 500 C at pressures up to 200 bars based on regression of the available literature data: U.S. Geol. Surv. Open-file Rept. 75 - 636, 31p.

_____ and _____, 1976, The volumetric properties of vapor saturated aqueous potassium chloride solutions from 0 to 400 based on a regression of the available literature data: U.S. Geol. Surv. Open-file Rept. 76-243, 6p.

_____ and Clynne, M.A., 1978, Pressure corrections for fluid inclusion homogenization temperatures: Abst. with Programs, Int. Assoc., Genesis Ore Deposits Symp., Alta, Utah, p.146.

_____ , _____ ,. and Brown, D.L., 1978, Freezing point depression of aqueous sodium chloride solutions: Econ. Geol., v. 73, p. 284 - 285.

Prezbindowski, D.R., and Larese, R.E., 1987, Experimental stretching of fluid inclusions in calcite - implication for diagenetic studies: Geology, v. 15, p. 333 -336.

Price, L.C., 1983, Geologic time as a parameter in organic metamorphism and vitrinite reflectance as an absolute paleogeothermometer: Jour. Petrol. Geol., v. 6, p. 5 - 38.

Price, R.A. and Hatcher, R.D. Jr., 1983, Tectonic significance of similarities in the evolution of the Alabama - Pennsylvania Appalachians and the Alberta - British Columbia Canadian Cordillera *in* Hatcher, R.D.Jr., Williams, H. and Zietz, I. (eds.) Contributions to the Tectonics and Geophysics of Mountain Chains: Geol. Soc. Amer. Memoir 158. p. 149 - 160.

Qing, H. and Mountjoy, E.W., 1994, Formation of coarsely crystalline, hydrothermal dolomite reservoirs in the Presqu'ile Barrier, Western Canada sedimentary basin: Amer. Assoc. Petrol. Geol. Bull., v. 78, p. 55 - 77.

Quinlan, G.M. and Beaumont, C., 1984, Appalachian thrusting, lithospheric flexure, and the Paleozoic stratigraphy of the eastern interior of North America: Can. Jour. Earth. Sci., v. 21, p. 973 - 996.

Radke, B.M. and Mathis, R.L., 1980, On the formation and occurrence of saddle dolomite: Jour. Sed. Petrol. v. 50, p. 1149 - 1168

Ramsayer, K. and Boles, J.R., 1986, Mixed-layer illite/smectite minerals in Tertiary sandstones and shales, San Joaquin Basin, California: Clays and Clay Minerals, v. 34, p. 115 - 124.

Reitan, P.H., Szekely, J. and Foster, B.P., 1970, Material emplacement models for dikes extending to the mantle: Trans. Amer. Geophys. Union (EOS), v. 51, p. 447.

Rickard, L.V., 1975, Correlation of Silurian and Devonian rocks in New York State: New York State Museum Map and Chart Series no. 24.

_____, 1991a, Sand, salt, and "scorpions", northern lowlands and Tug Hill plateau, p. 67 - 99, *in* Isachsen, Y.W., Landing, E. Lauber, L.V. (eds.) Geology of New York, a simplified account: NY State Museum, State Educ. Dept., 284p.

_____, 1991b, Oldest forests and deep seas, Erie lowlands and Allegheny plateau, p. 101 - 137, *in* Isachsen, Y.W., Landing, E. Lauber, L.V. (eds.) Geology of New York, a simplified account: NY State Museum, State Educ. Dept., 284p.

Roberson, H.E. and Lahann, R.W., 1981, Smectite to illite conversion rates, effects of solution chemistry: Clays and Clay Minerals, v. 29, p. 129 - 135.

Rodgers, J., 1967, Chronology of tectonic movements in the Appalachian region of eastern North America: Amer. Jour. Sci., v. 265, p. 408 - 427.

_____, 1970, The Tectonics of the Appalachians: Wiley- Interscience, New York, 271p.

Roedder, E., 1962, Studies of fluid inclusions I: low temperature application of dual-purpose freezing and heating stage: Econ. Geol. v. 57, p.1045 - 1061.

_____, 1984, Fluid Inclusions *in* Ribbe, P.H. (ed.) Reviews in Mineralogy: Mineral. Soc. Amer. 12, 644p.

_____ and Bodnar, R.J., 1980, Geologic pressure determinations from fluid inclusion studies: Ann. Rev. Earth Planet. Sci., v. 8, p. 263 - 301.

Rowley, D.B. and Kidd, W.S.F., 1981, Stratigraphic relationships and detrital composition of the medial Ordovician flysch of western New England: implications for the tectonic evolution of the Taconic Orogeny: Jour. Geology, v. 89, p. 199 - 218.

Sanders, J.E., 1963, Late Triassic tectonic history of north- eastern United States: Amer. Jour. Sci., v. 261, p. 501 - 524.

Savarese, M., Gray, L.M., and Brett, C. E., 1986, Faunal and lithologic cyclicity in the Centerfield Member (Middle Devonian, Hamilton Group) of western New York: a reinterpretation of depositional history, p. 32 - 56, *in* Brett, C.E. (ed.) Dynamic Stratigraphy and Depositional Environments of the Hamilton Group (Middle Devonian) in New York State, Part I: N.Y. State Mus., Dept. of Educ., 156p.

Schneidermann, N. and Harris, P.M. (eds.), 1985, Carbonate cements: Soc. Econ. Paleon. Mineral. Spec. Pub. 36, 379p.

Scholle, P.A. and Halley, R.B., 1985, Burial diagenesis: out of sight, out of mind!, p.309-334, *in* Schneiderman, N. and Harris, P.M. (eds.) Carbonate Cements: Soc. Econ. Paleon. Mineral. Spec. Pub., 36, 379p.

Sharp Jr., J.M., 1978, Energy and momentum transport model of the Ouachita Basin and its possible impact on formation of economic mineral deposit: Econ. Geol., v. 73, p. 1057 - 1068.

Shepherd, T.J., Rankin, A.H., and Alderton, D.H.M., 1985, A Practical Guide to Fluid Inclusion Studies: Blackie, Glasgow and London, 239p.

Shukla, V. and Baker, P.A. (eds.), 1988, Sedimentology and geochemistry of dolostone: Soc. Econ. Paleont. Mineral. Spec. Pub., 43, 266p.

Sloss, L.L., 1963, Sequences in the cratonic interior of North America: Geol. Soc. Amer. Bull., v. 74, p. 93 - 113.

_____, 1972, Synchrony of Phanerozoic sedimentary- tectonic events of North American craton and Russian platform: 24th Int. Geol. Cong., Montreal, Sect. 6, p. 24 - 32.

Smart, G. and Clayton, T., 1985, The progressive illitization of interstratified illite-smectite from Carboniferous sediments of northern England and its relationship to organic maturity indicators: Clay Minerals, v. 20, p. 455 - 466.

Sorby, H.C., 1858, On the microscopic structure of crystals, indicating the origin of minerals and rocks: Quart. Jour. Geol. Soc. Lond., v. 14, p. 453 - 500.

Stach, E., Mackowski, M.Th., Teichmuller, et al. 1982, Coal Petrology, 3rd Ed., Berlin-Stuttgart, Gebruder Borntraeger, 535p.

Stanley, R.S. and Ratcliffe, N.M., 1985, Tectonic synthesis of the Taconic orogeny in western New England: Geol. Soc. Amer. Bull., v. 96, p. 1227 - 1250.

Staplin, F.L., 1969, Sedimentary organic matter, organic metamorphism, and oil and gas occurrences: Can. Petrol. Geol. Bull., v. 17, p. 47 - 66.

_____, 1977, Interpretation of thermal history from color of particulate organic matter - a review: Palynology, v. 1, p. 9 - 18.

Stearn, R.G. and Reesman, A.L., 1986, Cambrian to Holocene burial history of Nashville Dome: Amer. Assoc. Petrol. Geol. Bull., v.70, p. 143-154.

Stevens, R.K., 1970, Cambro-Ordovician flysch sedimentation and tectonics in west Newfoundland and their possible bearing on a proto-Atlantic ocean, in Lajoie, J. (ed.) Flysch Sedimentology in North America: Geol. Assoc. Can. Spec. Pap. 7, p. 165-

Sweeney, J.J. and Burnham, A.K., 1990, Evaluation of a simple model of vitrinite reflectance based on chemical kinetics: Amer. Assoc. Petrol. Geol. Bull., v. 74, p. 1559 - 1570.

Tankard, A.J., 1986, On the depositional response of thrusting and lithospheric flexure: examples from the Appalachian and Rocky Mountain basins, in Allen, P.A. and Homewood, P. (eds.) Foreland Basins, Int. Assoc. Sedimentologists Spec. Pub. no. 8, p. 369 - 392.

Teichmuller, M., 1952, Die Anwendung des polierten Dunnschliffes bei der Mikroskopic von Kohlen and versteinerten Torfen, *in* Freund, H. (ed.), Handbuch der Mikroskopic in der Technik, Bd. 2, Teil I, p. 235 - 310. Umschau Verlag. Frankfurt am Main.

Thomas, W.A. and Schenk, P.E., 1988, Late Paleozoic sedimentation along the Appalachian orogen, *in* Harris, A.L. and Fettes, D.J. (eds.) The Caledonian - Appalachian Orogen: Geol. Soc. Spec. Pub. 38, p. 515 - 530.

Tilley, B.J., Nesbitt, B.E. and Longstaffe, F.J., 1989, Thermal history of Alberta Deep Basin: comparative study of fluid inclusions and vitrinite reflectance data: Amer. Assoc. Petrol. Geol. Bull., v. 73, p. 1206 - 1222.

Tillman, J.E. and Barnes, H.L., 1983, Deciphering fracturing and fluid migration histories in the northern Appalachian Basin: Amer. Assoc. Petrol. Geol. Bull., v. 67, p. 692 - 705.

Tissot, B.P., Pelet, R. and Ungerer, Ph., 1987, Thermal history of sedimentary basins, maturation indices and kinetics of oil and gas generation: Amer. Assoc. Petrol. Geol . Bull., v. 71, p. 1445 - 1466.

Treesh, M.I. and Friedman, G.M., 1974, Sabkha deposition of the Salina Group (Upper Silurian) of New York state, p. 35- 46, *in* Coogan, A.H. (ed.) Northern Ohio Geol. Soc. Sympos. on Salt, 4th Proc., v. 1, 530p.

Urschel, S.F. and Friedman, G.M., 1984, Paleodepth of burial of Lower Ordovician Beekmantown carbonates in New York State: Compass, Sigma Gamma Epsilon, v. 61, p. 205 - 215.

Van der Voo, R., 1979, Age of Alleghenian folding in the central Appalachians: Geology, v. 7, p. 297 - 298.

Velde, B., and Vasseur, G., 1992, Estimation of the diagenetic smectite to illite transformation in time-temperature space: Amer. Mineralogist, v. 77, p. 967 - 976.

Videtich, P.A., McLimans, R.K., Watson, H.K.S. and Nagy R.M., 1988, Depositional, diagenetic, thermal, and maturation histories of Cretaceous Mishrif Formation, Fateh Field, Dubai: Amer. Assoc. Petrol. Geol. Bull., v. 72, p. 1143-1159.

Vityk, M.O., Bodnar, R.J barometers: relation between pressure-temperature history and reequilibration morphology during crustal thickening: Geology, v. 22, p. 731 - 734.

Waples, D., 1980, Time and temperature in petroleum formation: application of Lopatin's method to petroleum exploration: Amer. Assoc. Petrol. Geol. Bull., v. 64, p. 916 - 926.

Weaver, C.E., 1978, Geothermal alteration of clay minerals and shales: diagenesis: Tech. Rept. ONW1-21, 176p.

_____ and Beck, K.C., 1971, Clay water diagenesis during burial: How mud becomes gneiss: Geol. Soc. Amer. Spec. Pap. no. 134, 96p.

_____ and Broekstra, B.R., 1984, Illite-mica, p.67-98, in Weaver, C.E. and associates (eds.) Shale-Slate Metamorphism in Southern Appalachians, Development in Petrology 10, Elsivier, 239p.

_____ and Wampler, J.M., 1970, K, Ar, illite burial: Geol. Soc. Amer. Bull., v. 81, p. 3423 - 3430.

_____, Eslinger, E.V. and Yeh, H.-W., 1984, Oxygen isotopes, p. 141-152, in Weaver, C.E. and associates (eds.) Shale-Slate Metamorphism in Southern Appalachians, Developments in Petrology 10, Elsivier, 239p.

Weiner, R.W., McLelland, J.M., Isachsen, Y.W. and Hall, L.M., 1984, Stratigraphy and structural geology of the Adirondack Mountains, New York: review and synthesis: Geol. Soc. Amer. Spec. Pap. 194, p. 1- 55.

Whitney, P.R. 1983, A three-stage model for the tectonic history of the Adirondack region, New York: Northeastern Geol., v. 5, p. 61 - 72.

_____, 1991, New mountains from old rocks, p. 23 - 44, in Isachsen, Y. W., Landing, E., Lauber, J.M. et al. (eds.) Geology of New York, a Simplified Account: NY State Mus., State Educ. Dept., 284p.

_____ and Davin, M.T.,1 987, Taconic deformation and metasomatism in Proterozoic rocks of the easternmost Adirondacks: Geology, v. 15, p. 500 - 503.

Williams, H., 1979, Appalachian orogen of Canada: Can. Jour. Earth Sci., v. 16, p. 792 - 807.

_____ and Hatcher, R.D.Jr., 1982, Suspect terranes and accretionary history of the Appalachian orogen: Geology, v. 10, p. 530 - 536.

Wood, G.H.Jr., Trexler, J.P., and Kehn, T.M., 1969, Geology of west-central part of the southern Anthracite field and adjoining areas, Pennsylvania: U.S. Geol. Surv. Prof. Pap. 602, 150p.

Zadins, Z.Z., 1984, Thrusting, folding and cleavage development in the Silurian -Devonian section of eastern New York: Abst. with Programs, Geol. Soc. Amer. northeastern sect. ann. meetg., p.73.

Zen, E-an, 1967, Time and space relationships of the Taconic allochthon and autochthon: Geol. Soc. Amer. Spec. Pap. 97, 107p.

_____, 1972, The Taconide zone and the Taconic orogeny in the western part of the northern Appalachian orogen: Geol. Soc. Amer. Spec. Pap. 135, 72p.

_____, 1983, Exotic terranes in the New England Appalachians - limits, candidates, and ages: a speculative essay in Hatcher, R.D., Williams, H. and Zietz.

I. (eds.) Contribution to the Tectonics and Geophysics of Mountain Chains: Geol. Soc. Amer. Memoir 158, p. 55 - 82.

Zenger, D.H., Dunham, J.B., and Ethington, R.L. (eds.), 1980, Concepts and models of dolomitization: Soc. Econ. Paleont. Mineral. Spec. Pub. 28, 379p.

Zhang, E. And Davis, A., 1993, Coalification patterns of the Pennsylvanian coal measures in the Appalachian foreland basin, western and south-central Pennsylvania: Geol. Soc. Amer. Bull., v. 105, p. 162 - 174.

SUBJECT INDEX

Springer-Verlag
and the Environment

We at Springer-Verlag firmly believe that an international science publisher has a special obligation to the environment, and our corporate policies consistently reflect this conviction.

We also expect our business partners – paper mills, printers, packaging manufacturers, etc. – to commit themselves to using environmentally friendly materials and production processes.

The paper in this book is made from low- or no-chlorine pulp and is acid free, in conformance with international standards for paper permanency.

Lecture Notes in Earth Sciences

Vol. 37: A. Armanini, G. Di Silvio (Eds.), Fluvial Hydraulics of Mountain Regions. X, 468 pages. 1991.

Vol. 38: W. Smykatz-Kloss, S. St. J. Warne, Thermal Analysis in the Geosciences. XII, 379 pages. 1991.

Vol. 39: S.-E. Hjelt, Pragmatic Inversion of Geophysical Data. IX, 262 pages. 1992.

Vol. 40: S. W. Petters, Regional Geology of Africa. XXIII, 722 pages. 1991.

Vol. 41: R. Pflug, J. W. Harbaugh (Eds.), Computer Graphics in Geology. XVII, 298 pages. 1992.

Vol. 42: A. Cendrero, G. Lüttig, F. Chr. Wolff (Eds.), Planning the Use of the Earth's Surface. IX, 556 pages. 1992.

Vol. 43: N. Clauer, S. Chaudhuri (Eds.), Isotopic Signatures and Sedimentary Records. VIII, 529 pages. 1992.

Vol. 44: D. A. Edwards, Turbidity Currents: Dynamics, Deposits and Reversals. XIII, 175 pages. 1993.

Vol. 45: A. G. Herrmann, B. Knipping, Waste Disposal and Evaporites. XII, 193 pages. 1993.

Vol. 46: G. Galli, Temporal and Spatial Patterns in Carbonate Platforms. IX, 325 pages. 1993.

Vol. 47: R. L. Littke, Deposition, Diagenesis and Weathering of Organic Matter-Rich Sediments. IX, 216 pages. 1993.

Vol. 48: B. R. Roberts, Water Management in Desert Environments. XVII, 337 pages. 1993.

Vol. 49: J. F. W. Negendank, B. Zolitschka (Eds.), Paleolimnology of European Maar Lakes. IX, 513 pages. 1993.

Vol. 50: R. Rummel, F. Sansò (Eds.), Satellite Altimetry in Geodesy and Oceanography. XII, 479 pages. 1993.

Vol. 51: W. Ricken, Sedimentation as a Three-Component System. XII, 211 pages. 1993.

Vol. 52: P. Ergenzinger, K.-H. Schmidt (Eds.), Dynamics and Geomorphology of Mountain Rivers. VIII, 326 pages. 1994.

Vol. 53: F. Scherbaum, Basic Concepts in Digital Signal Processing for Seismologists. X, 158 pages. 1994.

Vol. 54: J. J. P. Zijlstra, The Sedimentology of Chalk. IX, 194 pages. 1995.

Vol. 55: J. A. Scales, Theory of Seismic Imaging. XV, 291 pages. 1995.

Vol. 56: D. Müller, D. I. Groves, Potassic Igneous Rocks and Associated Gold-Copper Mineralization. XIII, 210 pages. 1995.

Vol. 57: E. Lallier-Vergès, N.-P. Tribovillard, P. Bertrand (Eds.), Organic Matter Accumulation. VIII, 187 pages. 1995.

Vol. 58: G. Sarwar, G. M. Friedman, Post-Devonian Sediment Cover over New York State. VIII, 113 pages. 1995.